Trends and Issues in

Roland Conrady · Martin Buck
Editors

Trends and Issues in Global Tourism 2009

 Springer

Editors

Professor Dr. Roland Conrady
University of Applied Sciences
  Worms
Erenburger Str. 19
67549 Worms
Germany
conrady@fh-worms.de

Dr. Martin Buck
Messe Berlin GmbH, ITB
Messedamm 22
14055 Berlin
Germany
buck@messe-berlin.de

ISBN 978-3-642-10074-1   e-ISBN 978-3-540-92199-8

DOI 10.1007/978-3-540-92199-8

ISSN 1868-0127   e-ISSN 1868-0135

Cover design: WMXDesign GmbH, Heidelberg

Printed on acid-free paper

9 8 7 6 5 4 3 2 1

springer.com

Preface

As other industries, the global travel and tourism industry has been facing immense challenges and highly visible upheaval since the beginning of the new millennium. The International Tourism Exchange ITB Berlin, the world's leading travel trade show, aims at pinpointing the most important challenges, identifying the trends and offering a platform to solve pressing problems. The ITB Convention Market Trends & Innovations has developed into a centre of excellence and a driving force for the global travel and tourism industry, generating a much needed information platform.

This compilation unites the highlights of the convention in articles prepared by renowned professionals and scientists from the industry. Readers may benefit from this comprehensive vision of the developments that are shaping the structure of the global tourism industry today and in the future. This book is indispensable for tourism and travel professionals as well as for academics and students analyzing current global tourism and travel trends.

The first chapter draws on current empirical data on travel behaviour of the world population to establish a foundation for this book. The World Travel Monitor conducts more than 1 m interviews per year in over 50 countries and represents the world's largest and most important travel survey. Rolf Freitag and Dennis Pyka provide insight concerning data of global travel behaviour and travel trends that are dominating the market. Philipp E. Boksberger, Kristian Sund and Markus R. Schuckert focus on one of the most important and enduring trend topics, the implications of socio-demographic change on the tourism branch. This paper discusses the possibilities and limits of today's tourism industry in regards to predicting future travel behaviour based on socio-demographic changes. Based on a written survey of German-speaking visitors at a destination in Switzerland, the results support the thesis of cohort-specific travel behaviour.

The second chapter deals with an issue that like no other deserves the special attention of the global travel and tourism industry: global climate change. First Eric Heymann und Philipp Ehmer present the results of a study from Deutsche Bank Research. The authors outline the ways in which the environmental-climatic and regulatory-market economy dimensions of climate change can affect the tourism industry. By using a scoring model, the authors differentiate tourist regions that can profit from climate change and those that are expected to be on the losing side. The forecast horizon of the Deutsche Bank Research investigation is 2030. Subsequently, Hansruedi Müller from the Research Institute for Leisure and Tourism (FIF) of the University of Berne examines the effects of climate change on Alps tourism. After presenting a detailed forecast of climate changes in Switzerland, in Austria and Germany, Müller analyses the opportunities and risks of climate change and provides recommended actions for handling climate change. Subsequently Christoph Kessler, Ralf Kadel und Laura Vogel grapple with another region strongly affected by climate change: Africa. As a long-haul destination and as a region that is already very hot today, the challenges for Africa are enormous. Innovative zero footprint examples show how climate-neutral tourism could look in Africa.

The third chapter is devoted to Aviation Management. Richard Klophaus addresses a topic that has already led to a drastic situation for many airlines in 2008: extremely high fuel prices. His paper examines the impact of rising fuel prices on future air traffic. Using route and carrier specific data the short-term impact of higher fuel prices on airline operating costs, passenger fares and demand for short-haul and long-haul services is analysed. Patrick S. Merten considers the future of passenger processes at airports and sheds light on changes from the perspective of passengers, airlines and airports. Mobile tickets, check-in kiosks, web and mobile check-in, and electronic and mobile boarding passes optimise the traditional airport process. Using information from expert and passenger surveys, the article reveals how a passenger process supported by technology could look at airports in the future.

The fourth chapter addresses issues of travel technology. It reflects the results of latest studies on the usage of consumer technologies relevant to travel and tourism. The PhoCusWright Consumer Technology Survey Second Edition focuses on Web 2.0 familiarity and usage patterns of U.S. travellers, on the online traveller shopping behaviour and the influence of consumer technologies on travel, on social media usage and influence. The usage of mobile devices is analysed as well.

The fifth chapter is devoted to the management of tourist destinations. In their article, Campbell C. Thomson and Harald Jahn from the European Investment Bank (EIB) analyse tourism strategies and policies in the southern and eastern countries bordering the Mediterranean (the so-called FEMIP countries). The article reviews the current position of the tourism sector across the region, identifying the strengths and weaknesses of each country, and whether the strategies being followed are compatible with the concept of Sustainable Tourism. Stefan Zimmermann and Tony Reeves examine the potentials and methods of movie tourism as an innovative and effective form for marketing of destinations.

The sixth chapter covers the topics of Marketing and Sales Management. Monika Echtermeyer looks at the source market of China as an example to probe the importance of brands for destination decisions of tourists. It appears that to a great extent Chinese tourists take their decisions depending on how far they encounter known brands at European destinations. In her article Echtermeyer also offers recommendations for management of destination and hotel brands catered to Chinese tourists. Klaus-Dieter Koch provides insight on the luxury market. He describes what opportunities the luxury market segment offers and how global luxury brands in tourism should be managed. Michael Vogel looks into the strongly growing market for cruises. He also analyses the international marketing strategies of the Royal Caribbean and Carnival cruise ship corporations.

The seventh chapter is directed towards Business Travel Management. Andreas Wilbers starts off with the current developments in the business travel sector and the resulting demands on Travel Management Companies. Jörg Martin turns his attention to the international aspects of Business Travel Management. Uwe Klapka investigates a particular aspect of internationality: multicultural interaction at meetings and events. Next, Andreas Krugmann discusses the topic of safety when travelling. The results of an empirical study and the implications for Business Travel Management complete this article.

The eighth chapter offers a perspective beyond the borders of the tourism industry. The cultural scientist Asfa-Wossen Asserate asks whether in view of new source markets for international tourism a clash of cultures exists and what the impacts would be. The philosopher Peter Sloterdijk rounds out the work with perspectives on the present economic global tourist scene.

This work could not have been achieved without the remarkable dedication on behalf of the authors, who for the most part have taken on executive positions in the tourism economy. Special thanks go to Pia Viehl from the Faculty of Tourism and Travel, University of Applied Sciences Worms. She tirelessly dedicated herself with extraordinary commitment, remarkable skill and well-founded expert knowledge to ensure timely publication of the work. In the process she never lost sight of our high quality standards and was thereby instrumental in the success of the work. Without her contribution, this work would not be in your hands now.

Our thanks also go to the team of highly competent and reliable translators of the Mainz/Germersheim University led by H.-J. Bopst, including K. Kleist, C. Obermaier, V. Piaggio, L. Russell, K. Schrader, V. Srinivasan, K. Stellrecht, T. Volkmer, R. Walker and M. Zivcic.

Frankfurt/Berlin, December 2008

Roland Conrady
University of Applied Sciences Worms

Martin Buck
Messe Berlin

Contents

2030: Alps Tourism in the Face of Climate Change 57

Hansruedi Müller

Zero Footprint – A Viable Concept for Climate-Friendly Tourism in Africa?.. 65

Ralph Kadel, Christoph Kessler, and Laura Vogel

Aviation Management

Kerosene's Price Impact on Air Travel Demand: A Cause-and-Effect Chain... 79

Richard Klophaus

Travel Technology

Destination Management

Film Tourism – Locations Are the New Stars 155

Stefan Zimmermann and Tony Reeves

Marketing and Sales Management

Brands as Destinations – The New Tourism Objective for Chinese Tourists 165

Monika Echtermeyer

Luxury Tourism – Insights into an Underserved Market Segment... 183

Klaus-Dieter Koch

Is Europe One Market or Many? The US Cruise Companies' Segmentation Problem 193

Michael Vogel

Business Travel Management

Beyond Tourism Industry's Boundaries: The Philosophers' View on Sociological Mega Trends

Authors

Asserate, Dr. Asfa-Wossen
   Niedenau 72
   60325 Frankfurt a. M.
   Germany

Boksberger, Prof. Dr. Philipp E.
   Director ITF
   Institute of Tourism and Leisure Research (ITF)
   University of Applied Sciences HTW Chur
   Comercialstrasse 22
   7000 Chur
   Switzerland
   philipp.boksberger@fh-htwchur.ch

Buck, Dr. Martin
   Director
   Messe Berlin GmbH
   Competence Centre Travel & Logistics
   Messedamm 22
   14055 Berlin
   Germany
   buck@messe-berlin.de

Conrady, Prof. Dr. Roland
   University of Applied Sciences Worms
   Department of Tourism and Travel
   Erenburger Str. 19
   67549 Worms
   Germany
   conrady@fh-worms.de

Echtermeyer, Prof. Dr. Monika
    Specialist for Outbound-Tourism China
    International University of Applied Sciences Bad Honnef-Bonn
    Campus Bad Reichenhall
    Mülheimerstr. 38
    53604 Bad Honnef
    Germany
    m.echtermeyer@fh-bad-honnef.de

Ehmer, Philipp
    Economist
    Deutsche Bank AG
    Deutsche Bank Research
    Theodor-Heuss-Allee 70
    60486 Frankfurt a. M.
    Germany
    philipp.ehmer@db.com

Freitag, Rolf
    CEO
    IPK International
    Gottfried-Keller-Straße 20
    81245 Munich
    Germany
    freitag@ipkinternational.com

Jahn, Dr. Harald
    Head of Services, Agroindustry and SMEs
    Projects Department
    European Investment Bank
    100, boulevard Konrad Adenauer
    2950 Luxembourg
    Luxembourg
    Jahn@eib.org

Heymann, Eric
    Senior Economist
    Deutsche Bank AG
    Deutsche Bank Research
    Theodor-Heuss-Allee 70
    60486 Frankfurt a. M.
    Germany
    eric.heymann@db.com

Kadel, Dr. Ralph
    Senior Project Manager
    Sub Sahara Africa
    KfW Development Bank
    Palmengartenstr. 5-9
    60325 Frankfurt a. M.
    Germany
    Ralph.Kadel@kfw.de

Kessler, Dr. Christoph
    Director
    KfW Development Bank
    Palmengartenstr.5-9
    60325 Frankfurt a. M.
    Germany
    christoph.kessler@kfw.de

Klapka, Uwe
    Executive Director Germany
    Meeting Professionals International (MPI)
    Crellestr. 21
    10827 Berlin
    Germany
    uklapka@mpiweb.org

Klophaus, Prof. Dr. Richard
    Centre for Aviation Law and Business
    University of Applied Sciences Trier
    Postfach 1380
    55761 Birkenfeld
    Germany
    klophaus@umwelt-campus.de

Koch, Klaus-Dieter
    Managing Director
    Brand:Trust GmbH
    Findelgasse 10
    90402 Nürnberg
    Germany
    kdk@brand-trust.de

Krugmann, Andreas
   Manager Business Travel Industry
   Mondial Assistance Group
   ELVIA Reiseversicherungs-Gesellschaft AG
   Ludmillastr. 26
   81543 Munich
   Germany
   andreas.krugmann@elvia.de

Martin, Jörg
   Managing Director
   CTC Corporate Travel Consulting
   Hamsterweg 22b
   65307 Bad Schwalbach
   Germany
   martin@ctcnet.de

Merten, Patrick S.
   Aviation Research
   Auf dem Feldele 4a
   79227 Schallstadt
   Germany
   patrick.merten@aviation-research.de

Müller, Prof. Dr. Hansruedi
   Director
   Research Institute for Leisure and Tourism (FIF)
   University Bern
   Schanzeneckstrasse 1
   3001 Bern
   Switzerland
   hansruedi.mueller@fif.unibe.ch

Pyka, Dennis
   IPK International
   Gottfried-Keller-Straße 20
   81245 Munich
   Germany
   pyka@ipkinternational.com

Reeves, Tony
   The Worldwide Guide to Movie Locations
   4 Scholefield Road
   London N19 3EX
   UK

Schetzina, Cathy
  PhoCusWright
  Market Research – Industry Intelligence
  1 Route 37 East, Suite 200
  Sherman, CT 06784-1430
  USA

Schuckert, Dr. Markus R.
  Vice Director ITF
  Institute of Tourism and Leisure Research (ITF)
  University of Applied Sciences HTW Chur
  Comercialstrasse 22
  7000 Chur
  Switzerland
  markus.schuckert@fh-htwchur.ch

Sloterdijk, Prof. Dr. Peter
  Philosopher and President Karlsruhe School of Design
  Karlsruhe School of Design
  Lorenzstr. 15
  76135 Karlsruhe
  Germany
  psloterd@hfg-karlsruhe.de

Sund, Dr. Kristian J.
  Managing Director, Executive Master Programs
  College of Management of Technology
  Ecole Polytechnique Federale de Lausanne (EPFL)
  BAC 001 (Bassenges)
  1015 Lausanne
  Switzerland
  kristian.sund@epfl.ch

Thomson, Eur.Ing Campbell C.
  Advisor, Services, Agroindustry and SMEs
  Projects Department
  European Investment Bank
  100, boulevard Konrad Adenauer
  2950 Luxembourg
  Luxembourg
  Thomson@eib.org

Vogel, Prof. Dr. Michael P.
   Degree programme Cruise Industry Management
   Institute for Maritime Tourism
   University of Applied Sciences Bremerhaven
   An der Karlstadt 8
   27568 Bremerhaven
   Germany
   mvogel@hs-bremerhaven.de

Vogel, Laura
   Sciences Po Paris
   27 rue Saint-Guillaume
   75337 Paris Cedex 07
   France
   Laura.Vogel@sciences-po.org

Wilbers, Prof. Dr. Andreas
   University of Applied Sciences Worms
   Department of Tourism and Travel
   Erenburger Str. 19
   67549 Worms
   Germany
   wilbers@fh-worms.de

Zimmermann, Dr. Stefan
   Institute of Geography
   Johannes Gutenberg-University Mainz
   P.O. Box
   55099 Mainz
   Germany
   s.zimmermann@geo.uni-mainz.de

Key Figures and Forecasts in the Global Tourism Industry

Global Tourism in 2008 and Beyond – World Travel Monitor's Basic Figures

Rolf Freitag and Dennis Pyka

1 Introduction

This *Report* is based primarily on the 2007 results of IPK International's World Travel Monitor – the continuous tourism monitoring system, which was set up in 1988. IPK now undertakes more than half a million representative interviews a year in 57 of the world's major outbound travel markets – 36 in Europe and 21 in the rest of the world – representing an estimated 90% of world outbound travel.

The interviews – more than 5 million of which have now been conducted since 1988 – are designed to be comparable from one year and from one market to another, and to yield information on market volumes and sales turnover, destinations, travel behaviour, motivation and satisfaction, travellers and target groups, recent tourism trends, and short- to medium-term forecasts.

The report focuses as usual on world and European trends, looking in more detail at the German travel market, domestic and outbound.

Once again, both the World Travel Monitor results and the other sources of information presented in this report reflect the healthy state of the global travel and tourism industry in 2007.

2 Overview of World Tourism in 2007

2.1 Global Trends

2007 was the fourth year in succession of very rapid growth for world tourism, according to IPK International's World Travel Monitor, the results of which correlate with those identified by the World Tourism Organization (UNWTO).

While one could argue – as the following graph seems to suggest – that the industry is still making up ground lost in the first, crisis-strewn, years of the new century, it is clear that, in line with global economic growth trends, travel and tourism has not had it so good for a very long time. Growth over the period 2004-07 was much higher than that in the second half of the 1990s, prior to the crises of this decade.

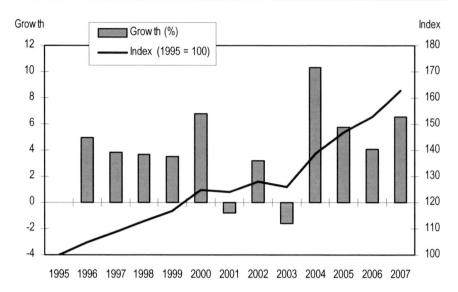

Fig. 1. World outbound tourism performance, 1995-2008. Source: World Travel Monitor, IPK International

The World Travel Monitor shows that in 2007, international overnight trips (of a minimum one night abroad) by travellers aged 15 and over increased by 6.3% worldwide – well above the long-term annual average of 4.3% – to 680 million. Overnight volume was also up 6% to 7.2 billion. (By way of interest, UNWTO reports about 900 million arrivals, but these include same-day trips and transfers for some markets, children under the age of 15, and cumulative arrivals in several countries visited on one trip.)

According to its own specific measure – which includes total spending related to a trip, both during and before the trip – the World Travel Monitor suggests that expenditure rose by 7% to €753 billion last year, equivalent to €117 per night and €1,150 per outbound trip.

It was not that there were no problems in 2007:

- The conflicts in Iraq, Afghanistan, Pakistan, Palestine, Sri Lanka and elsewhere constantly threatened to spill further afield, especially in the form of mindless terrorism.

- The persistent decline in the value of the US dollar, especially against the euro and sterling, affected the purchasing power abroad of many potential travellers and had a negative impact on the profitability of tourism operations, penalising especially those which are priced in advance in dollars and which are not large and sophisticated enough to be able to hedge their exposure.

- The first concerns about a potential recession in the USA took hold of the imaginations of both consumers and business.

- International oil prices continued to rise, approaching the US$100 a barrel mark by year end, and bringing further increases in airline fuel surcharges.

- There was the usual crop of weather-related problems, including a short and snowless winter in many Alpine resorts, floods in the UK, heatwaves and forest fires in Greece and Italy, and the panoply associated with an El Niño event – droughts in the Eastern USA, Southern Africa and Australasia, and floods in the Western Americas and Southern/Eastern Asia, for instance.

However, these challenges were not serious enough to undermine the growth in global travel demand, which was driven in large part by the sustained world economic growth. Among the key factors were the continuation of the unprecedented growth rates in emerging Asian markets; the surge in commodity prices that brought prosperity to Latin America, Russia and even some parts of Africa; the ongoing boom in East European markets as they become more integrated in the European Union; and signs of more robust growth in the more westerly members of the European Union. Of particular interest to many participants in the ITB Berlin Convention was the economic recovery in Germany and neighbouring countries.

Associated with this prosperity is a fundamental change in attitudes brought about by more widespread affluence, globalisation, the internet, mobile phones and cheap air travel. Two generations ago leisure travel abroad was limited to the cultural elites of a handful of countries. One generation ago it had spread to the elites worldwide, and to the middle classes of a few countries. Today, huge numbers of people, in countries which until recently were 'out of the loop', are becoming aware of, and attracted to, foreign destinations, and the possibility of travel.

As everyone knows, the world's media have a fondness for predicting apocalyptic comeuppances for the tourism industry – as was the case, for example, after 9/11, the SARS epidemic and the outbreak of bird flu. Last year, the media went even further in their victimisation of the travel and tourism industry, awarding air travel the title of World Climate Killer Number 1. Nevertheless, there has so far been little sign of any serious impact of this on travel demand – as demonstrated by the World Travel Monitor's results for 2007.

2.2   Outbound Performance

The economic prosperity which has been driving the growth in world tourism has been centred on emerging countries, especially in Asia and Europe, as the following graph shows. It is difficult to understate the rapidity with which purchasing power has been growing in the more successful emerging economies.

This shift in focus is expected to persist, partly because of the huge amount of ground which the more heavily populated emerging countries still have to make

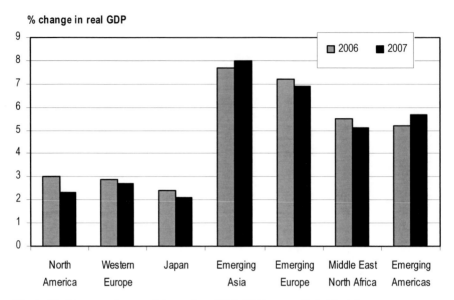

Fig. 2. World economic growth by region, 2006-07. Source: Global Insight

up, and partly because the terms of trade seem to have shifted – at least for the time being – in favour of commodities' producers. To put it another way, international oil, minerals and foodstuff prices are high, and are expected to remain high. Nevertheless, Europe contributed the lion's share of the increase in world tourism in 2007 – 18 million of the 40 million extra outbound trips. Why should that be?

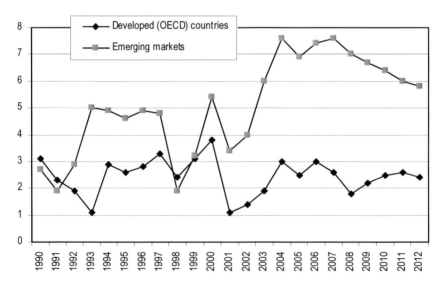

Fig. 3. Increase in global purchasing power, 1990-2012 (%). Source: Global Insight

Table 1. Growth in outbound tourism by region, 2007

Origin	Total (mn trips)	Growth (%)	Increase (mn trips)
Europe	407	5	18
Asia Pacific	152	10	15
Americas	107	6	6
Africa	14	5	1
Total	**680**	**6**	**40**

Note: Asia Pacific includes West Asia/Middle East and Central Asia

Source: World Travel Monitor, IPK International

First, international tourism within, and out of, Europe is on such a large scale that a relatively modest growth rate still yields the largest number of additional trips. Second, things are not quite as they seem at first glance: much of the growth in European outbound trips in 2007 was contributed by the 'emerging' countries of Central and Eastern Europe, just as most of the growth in American trips was contributed by Latin America, not by the developed countries of North America.

Similarly, Europe remains an extremely important source of spending on travel since expenditure is very strongly related to buying power. The Swiss, for example, spend twelve times more than the Poles on outbound travel per capita, as illustrated by the following graph.

The graph also highlights the scope for immense increases in travel spending by emerging markets with substantial populations but, as yet, very modest buying power, not just in Europe, but all over the world.

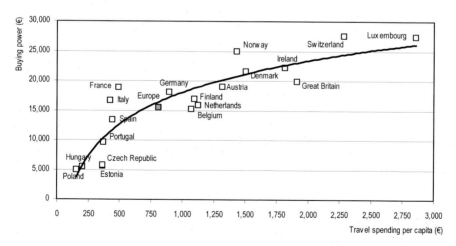

Fig. 4. Relationship of outbound travel spending to buying power in Europe, 2007. Source: World Travel Monitor, IPK International; GfK buying power analysis

2.3   Inbound Performance

Analysis of world outbound travel by destinations visited shows that Asia Pacific, Africa and the Americas achieved very substantial growth in arrivals in 2007, but the dominance of 'destination Europe' in the overall count, which saw an increase of 'only' 4% in arrivals, pulled down the global average to 6%.

% change in international arrivals

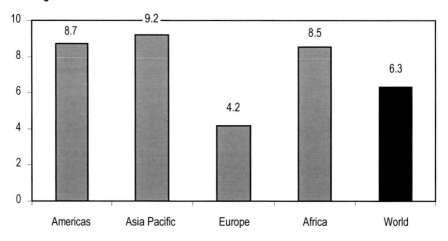

Note: Asia Pacific includes West Asia/Middle East and Central Asia

Fig. 5. Growth in international inbound tourism by region, 2007. Source: World Travel Monitor, IPK International

According to the World Travel Monitor, the Northern, Western and Southwest-Mediterranean sub-regions of Europe all achieved increases of 2-4% in both arrivals and visitor spending. The Southeast-Mediterranean region performed far better, with an increase of 9% in arrivals and 8% in spending. Central and Eastern Europe saw a growth of only 2% in arrivals, but 18% in spending, the World Travel Monitor suggests – which probably had a lot to do with increased inflation coupled with rising exchange rates in some countries (notably Russia).

Tourism in the Americas did far better in 2007 than in 2006. Arrivals were up nearly 9% (compared with just 2% in 2006) and spending 12%. This was principally due to the recovery in inbound travel to the USA, thanks in part to the weak US dollar. Conversely, with a strong Canadian dollar and stricter formalities on the US border, there was a decline in arrivals in Canada. Overall, arrivals in North America rose by 8% and visitor expenditure grew by 12%.

Elsewhere in the Americas, tourism to Central America remained relatively strong, with increases of 7% in arrivals and 12% in spending, and that to the Caribbean weak, with a decline of 2% in arrivals and an increase of only 4% in (dollar-denominated) expenditure. South America did even better than in 2006, with exceptional increases of 11% in arrivals and 20% in spending.

Given the economic problems of the continent, the acceleration of inbound tourism to Africa in 2006 and 2007 exceeded expectations. Overall, arrivals were up 8.5%, with good performances from both arrivals and spending for North Africa, sub-Saharan Africa and the Indian Ocean Islands.

Overnight tourism to the vast Asia Pacific region continues its breathtaking rise: arrivals were up over 9% and expenditure 13%, according to the World Travel Monitor. South Asia (ie the Indian subcontinent), Southeast Asia and Australasia all saw 'relatively modest' increases of around 7% in arrivals and 12% in spending. The stronger performances were put in by Northeast Asia, with rises of 10% and 13%, and West Asia (a sub-region which IPK defines to include the Middle East and Central Asia), with rises of 15% and 16%.

(Given the different definitions and methodologies, these figures cannot be readily compared with data from the Pacific Asia Travel Association presented later in this report. For instance, PATA not only defines the geographical region differently, but its definition of arrivals also includes same-day trips.)

3   Overview of European Tourism in 2007

3.1   Overall Travel Demand

In 2007 Europeans made 407 million trips abroad of a minimum one night's stay, according to IPK International's European Travel Monitor. These generated 3.9 billion overnights and €381 billion in spending.

While outbound trip and overnight volume both increased overall by 5% – with the result that there was no change in average length of trip at just under 10 nights –

Table 2. European outbound travel, 2007

	2007	% change on 2006
Trips (mn)	407	5
1-3 nights long	99	–2
4+ nights long	308	7
Overnights (mn)	3,900	5
Average length of trip (nights)	10	0
Spending (€ bn)	381	7
Spending per trip (€)	938	2
Spending per night (€)	99	2

Source: European Travel Monitor, IPK International

travel expenditure rose by an even more impressive 7%, generating a 2% increase in spending per trip, to €938, and in spending per night, to €99.

It should be noted that this was slightly less than average inflation in Europe and considerably less than average inflation in most other destinations. So the apparent improvement in 2007, in fact, did little more than keep pace with the increase in inflation.

The trend towards shorter but more frequent trips, which seemed to be so well entrenched over the last few years, was reversed in 2007. Trips of four nights and longer increased by 7% in volume, now accounting for a 76% share, while the number of trips of one to three nights actually fell by 2%. This is attributable, at least in part, to a decline in the number of secondary trips taken by mature European markets – whether for economic reasons or, indeed, in line with efforts by individuals to reduce their carbon emissions. But there is no clear evidence of the latter.

As will be discussed later in this report, in the section on 'Intra- versus Intra-regional Travel', demand for long-haul destinations increased by almost twice the rate as that for short-haul, or intra-regional, points. But intra-European travel still accounts for a massive 86% share of total European outbound trip volume.

Holidays account for more than two thirds of total European outbound trips (69%) while business travel, which had lost share in previous years but which started a sustained recovered in 2005, recorded a third consecutive year of stronger than average growth in 2007, taking its share to 15% – up from just 11% in 2004. Visits to friends and relations (VFR) and other leisure trips accounted for some 16% of total trips, the same as in 2006.

Table 3. Purpose of travel by Europeans, 2007

Purpose	Trips (mn)	% market share	% change 2007/06
Holiday	280	69	5
VFR and other leisure	65	16	0
Business	62	15	9
Total trips	**407**	**100**	**5**

Source: European Travel Monitor, IPK International

For the second year running, the last four months of the year showed the strongest growth in outbound trips from European countries. It is not clear whether these figures simply reflect seasonal factors, or an underlying strengthening of demand through the year – which would augur well for 2008. The sustained economic prosperity in Europe and the strength of (and confidence in) the euro suggest a strengthening of demand as the year wore on, but this was also a period of increasing international financial and economic uncertainty.

Table 4. European outbound travel growth by season, 2007

Season	% annual change in no. of trips
Jan-Apr	2
May-Aug	5
Sep-Dec	7

Source: European Travel Monitor, IPK International

Most likely, the recovery of long-haul travel had a lot to do with the healthy increase in demand from the month of October – the peak season for long-haul trips – which would have boosted overall growth in the last four months.

As far as holiday travel is concerned, there were strong performances in 2007 from touring holidays (+15%) and holidays in the countryside (+9%). Meanwhile, the largest segment, sun & beach holidays – which account for over 40% of total holiday trip volume – grew by 6%. This was the third successive year of less dynamic increases for sun & beach (they were up 4% in 2006 and 3% in 2005), but they remain the largest component of the European holiday travel market by a very wide margin.

City breaks, which for years had been the main beneficiaries of the boom in low-cost/low-fare fights, also seem to be growing more gently as demand approaches satisfaction levels in key markets – they were up 5% in 2006 and 5% again in 2007, to around 46 million trips.

The slower growth for city breaks was also due in part to the smaller number of mega-events in Europe in 2007, compared with 2006, when demand was boosted significantly by the FIFA Football World Cup in Germany, the year-long celebrations of Rembrandt's 400th anniversary in the Netherlands, and the commemoration of the 250th anniversary of Mozart's birth in several European countries, but primarily in Austria and Germany.

Table 5. European outbound holiday travel growth for selected segments, 2007

Type of trip	% annual change in no. of trips
Touring	15
Countryside	9
Sun & beach	6
City breaks	5

Source: European Travel Monitor, IPK International

3.2   Major Source Markets and Destinations

Germany, the UK, France and Italy continue to be Europe's leading outbound travel markets. But the Netherlands, Switzerland and, to a lesser extent, Belgium (because of their citizens' high propensities to travel and because of the Swiss high buying power) are also important markets, in spite of their smaller populations.

However, travel out of Spain and Russia has been rising the most rapidly in recent years, and 2007 was no exception. Both source countries have now overtaken Switzerland and Belgium, moving into sixth and seventh places in the ranking of the largest European markets. Italy has also put in relatively strong growth rates since 2005-06 and Norway, Ireland and Sweden also recorded above average increases last year.

Together, the top eight markets accounted for very nearly two thirds of total trip volume in 2007.

Table 6. Leading European outbound travel markets, 2007

Rank	Market	Trips (mn)	% change on 2006
1	Germany	75.9	2
2	UK	63.5	2
3	France	31.2	4
4	Italy	22.4	6
5	Netherlands	22.3	4
6	Russia	18.5	16
7	Spain	17.7	15
8	Switzerland	14.6	3

Source: European Travel Monitor, IPK International

The following graph shows the great range in travel spending per capita among European markets. As already indicated, the Swiss spend twelve times more per capita than the Poles. It is interesting to note the relatively low spending per capita on outbound travel among the French, Spanish and Italians, but this is also due in part to a high share of short cross-border trips. Conversely, the British, Dutch and Belgians are heavy spenders on outbound travel.

The table also highlights the huge scope for growth in the poorer countries of Europe (mainly but not entirely in Central and Eastern Europe) as their buying power – a combination of GDP, disposable incomes and price levels – converges with that in the richer countries.

Air travel continued to increase its share of the European outbound travel market in 2007. Trips by air increased by 8%, as against increases of 3% by ship, 2%

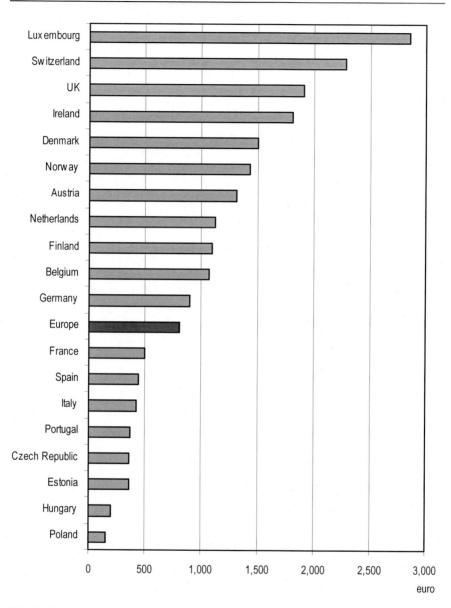

Fig. 6. Outbound travel expenditure per person in selected European countries, 2007. Source: European Travel Monitor, IPK International

by bus/coach and 1% by car. However, trips by train were also up sharply (+10%), reflecting the extension of high-speed rail services – notably the inauguration of the first phase of the TGV Est in France, reducing rail travel times to eastern France, Luxembourg, Germany and Switzerland – and maybe also a first consumer reaction to the climate change debate.

The growth in low-fare air travel in Europe continues (a low-fare flight, by IPK International's definition, is one costing less than €150 for a return trip), but the rate of growth has moderated. Low-fare airline trips accounted for 35% of total airline-based trips in Europe – unchanged over 2006's level. It is possible that this slower growth in low-fare airline travel is also partly due to the rise in fuel surcharges – ie some flights will have been transferred to the higher-cost category by the surcharges – but, as for the demand in city breaks, the main reason for the trend is the market's natural lifecycle.

The top eight destinations for European travellers are all within Europe, although the USA – the leading long-haul destination – follows fairly closely behind the leaders. The ranking of the top eight remained unchanged in 2007. Among

Table 7. European outbound travel – low-fare versus traditional flights, 2005-06

Type of flights	% growth 2007/06	% market share 2005	2006	2007
Traditional	6	67	65	65
Low-fare	13	33	35	35
All flights	8	100	100	100

Source: European Travel Monitor, IPK International

Table 8. Leading destinations of European outbound travellers, 2007

Rank	Destination	Trips (mn)	% market share	% change 2007/06
1	Spain	52	13	1
2	France	42	10	2
3	Germany	35	9	4
4	Italy	33	8	7
5	UK	21	5	2
6	Austria	21	5	2
7	Turkey	17	4	16
8	Greece	14	3	9

Source: European Travel Monitor, IPK International

these top eight destinations, the best growth was for Turkey (+16%) – which compensated for the decline the previous year – Greece (+9%) and Italy (+7%). The boom in travel to Central and Eastern appears to have subsided as total trips to the sub-region grew by only 2% in 2007. North Africa, meanwhile, attracted a 12% increase in visits by Europeans.

Only 14% of outbound trips from European countries were to long-haul destinations (ie destinations outside Europe and the Mediterranean basin). Trips to the Americas, the most important destination region outside Europe, increased by 8.5% – with North America up 12% as against +7% for Latin America and +2% for the Caribbean – while Asia attracted an 8% rise in trips from Europe, and the Middle East a 15% increase.

For the first time, IPK International has published information on the most popular cities visited by Europeans. Paris remains the favourite, attracting 11.5 million European visits, ahead of London with 11.2 million. (London, nevertheless, attracts a greater volume of non-Europeans.) And Berlin has climbed into third position in the overall ranking with 5.3 million European visits in 2007, compared with Vienna's 5.2 million.

Munich overtook Prague in the ranking, moving into seventh place, while Istanbul now shares tenth place with New York. Cities showing the best growth in terms of European visits – all recording double-digit increases – were New York, Berlin, Munich, Vienna and Istanbul.

Table 9. Europeans' favourite city destinations, 2007

City	No. of visits (mn)
Paris	11.5
London	11.2
Berlin	5.3
Vienna	5.2
Rome	4.5
Amsterdam	3.8
Munich	3.7
Prague	3.6
Madrid	3.4
New York	3.3
Istanbul	3.3

Source: European Travel Monitor, IPK International

3.3 Booking Patterns

As far as the organisation of travel was concerned, traditional travel agencies continued to lose ground in 2007 in favour of online travel – a trend that is evident all over the world. Some 50% of European outbound trips (up from 45% in 2006) involved the internet, 36% for actual booking (+14%) and 14% for just 'looking' (+6%) – gathering information relevant to the trip. Trips booked through traditional distribution channels fell by 1%, but lost five percentage points in terms of share to 50%.

All this is part of the continued growth in use of the internet for travel organisation – a trend which is maturing, and in which the headline growth rates are therefore moderating.

Table 10. European online travel, 2006-07

| | % of holidaymakers | | % increase |
	2006	2007	2007/06
Use of the internet	45	50	12
Bookers	32	36	14
Lookers[a]	13	14	6
No internet	55	50	−1

[a] Use of the internet to research travel options, but not for booking

Source: European Travel Monitor, IPK International

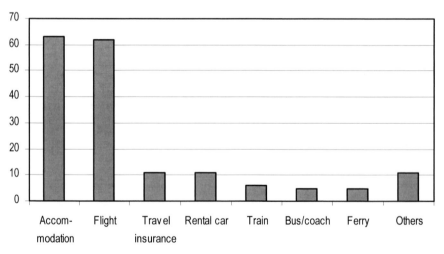

[a] Multiple responses possible

Fig. 7. Online bookings by European outbound travellers, 2007 (% of online 'bookers' booking each item online[a]). Source: European Travel Monitor, IPK International

The incidence of internet bookings among the smaller, less mature European markets is clearly much lower, but it is growing fast – much faster, in fact, than online travel 'looking'.

The total number of trips involving online booking in 2007 was about 141 million. Of these, around two thirds involved some kind of booking for accommodation and flights. Other products are much less likely to be booked online, either because the facility for such bookings is less widely available, or because Europeans are more nervous of booking these products through the internet.

4   German Travel & Tourism

4.1   Domestic and Outbound Trends

IPK International has been monitoring the German travel market since 1988, progressively increasing its scope and coverage. A total of some 24,000 interviews are now conducted every year for the German Travel Monitor (Deutscher Reisemonitor), making it the largest single travel survey in the country, with interviews carried out on a weekly basis.

According to the German Travel Monitor, Germans made a total of 297 million trips (domestic and international, of one night or longer) in 2007, an increase of 4% over 2006 (and a significant improvement on 2006's 1% increase over 2005). Overnight volume increased by 3% to 1.5 billion. However, as last year, Germans' spending on these trips rose by only 1%, to €124 billion – not enough to keep pace with inflation. One of the main reasons for this, of course, was the three percentage point increase in VAT in January 2007.

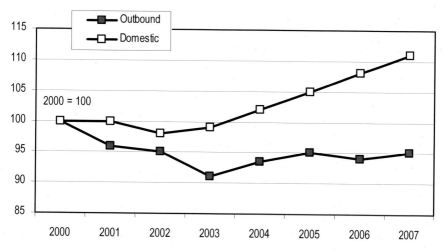

Fig. 8. Trends in German domestic and outbound travel, 2000-07. Source: German Travel Monitor, IPK International

Table 11. German domestic and outbound travel, 2007

	Total	% change on 2006
Trips (mn)		
Domestic	221	4
Outbound	76	2
Total	297	4
Nights (mn)		
Domestic	785	6
Outbound	698	0
Total	1,483	3
Spending (€ bn)		
Domestic	62	5
Abroad	62	−2
Total	124	1

Source: German Travel Monitor, IPK International

As many as 74% of these trips were domestic; only 26% were to foreign destinations. Since 2000 domestic trips have generally been increasing their share of the total market, and this trend continued in 2007. In fact, outbound trip volume was still 5% down on 2000's level in 2007, while domestic trips were 11% more numerous.

The total number of domestic trips increased by 4% to 221 million in 2007. With a 6% increase in nights spent on these trips, the average length of domestic trips increased slightly to 3.5 nights, reversing a significant decline in 2006. Spending on domestic trips rose by 5% to €62 billion (an average of €290 per trip).

Within the 4% increase in number of domestic trips, holidays were up 3%, VFR and other private trips up 2%, and business trips up 10% – the second year in succession of strong growth in business travel. The patterns of holiday travel returned to normality after the 2006 FIFA World Cup – trips to public events were sharply down, while tours were up 30%, city breaks up 32%, health and wellness trips up 24% and mountain holidays up 3%.

German outbound trips increased by 2% to 76 million in 2007. They lasted an average of 9.2 nights in 2007 (slightly less than in 2006), with an average spend of around €815 per trip and around €88 per night (both also down on 2006).

The demand for foreign holidays has been weak for several years, with declines of 2% in 2006 and 1% in 2007. Other private trips, including visits to friends and relations (VFR), recovered by 3%, after a 3% decline the previous year. Outbound travel demand is therefore being sustained only by business travel, which increased by 12% in 2007 in terms of trips, after a 6% increase in 2006.

Table 12. German outbound travel by purpose of trip, 2007

Purpose	No. of trips (mn)	% market share	% change on 2006
Holidays	52	69	−1
Other private reasons	15	19	3
Business	9	12	12

Source: German Travel Monitor, IPK International

Table 13. Leading destinations of German outbound holiday travellers, 2007

Rank	Destination	% market share	% change on 2006
1	Spain	17	−3
2	Austria	15	−1
3	Italy	14	7
4	France	7	−5
5	Turkey	6	10
6	Netherlands	5	7
7	Switzerland	4	5
	Top 7 destinations	**68**	**2**

Source: European Travel Monitor, IPK International

Table 14. German online travel, 2006-07

	% of holidaymakers		% increase
	2006	**2007**	**2007/06**
Use of the internet	44	49	6
Bookers	27	31	11
Lookers[a]	16	18	−3
No internet	56	51	−6

[a] Use of the internet to research travel options, but not for booking

Source: German Travel Monitor, IPK International

The top seven holiday destinations for German outbound travellers – which account for slightly over two thirds of total trips – are all within Europe. Demand for the top four – Spain, Austria, Italy and France – has tended to slip in recent years, but Italy staged a partial recovery in 2007. Turkey also suffered a heavy decline in 2006, but recovered strongly in 2007 – although not sufficiently to regain its 2005 market share.

The proportion of Germans who use the internet to organise their travel (44%) is closely comparable to the European average, albeit with rather more 'lookers' and fewer 'bookers'. The German and European figures both rose by five percentage points in 2007.

5   European Travel Trends in 2008

5.1   Overview of Main Trends in 2008

In 2007, according to IPK's European Travel Monitor, European adults aged 15 years and over made 407 million trips abroad. This represented an increase of 5% over 2006, and was followed by a 3% increase in the first eight months of 2008.

Of the 407 million trips made in 2007, 280 million (69%) were for holidays, 65 million (16%) were visits to friends and/or relations (VFR) and travel for other leisure purposes, and 62 million (15%) for business. Business travel, which had lost market share in previous years, has rebounded over the last three years, with growth rates consistently higher than the averages for all trips. Although the trend is now expected to reverse in the last few months of 2008 and in 2009 due to the economic slowdown, it was sustained in the first eight months of 2008, with business trip volume up a further 8%.

Table 15. European outbound travel, 2007-08

	2007	% change 2007/06	% change Jan-Aug 2008/07[b]
Trips[a] (mn)	407	5	3
Short trips (1-3 nights long)	99	−2	8
Long trips (4+ nights)	308	7	1
Holiday	280	5	4
VFR and other leisure	65	0	−5
Business	62	9	8
Overnights (mn)	3,900	5	3
Average length of stay (nights)	10	0	0
Spending (€ bn)	381	7	6
Spending per trip (€)	938	2	3
Spending per night (€)	99	2	2

[a] Trips made by adults aged 15 years and over   [b] Based on trends in the first eight months of 2008 from the leading 11 source markets, which account for 65% of European outbound trip volume

Source: IPK International's European Travel Monitor

Holiday travel increased by 4% from January through August – much the same rate of growth as in the same period of 2007 – but most of this growth was concentrated in the first part of the year. Holiday travel was up 6% in the first four months (thanks mainly to a better winter sports season), but growth slowed to only 1% in May to August, according to the European Travel Monitor. And, with the exception of isolated markets, no growth is expected in the last four months of the year.

In contrast to business and holiday travel, VFR and other leisure trips (undertaken mainly for educational, medical and/or religious purposes) declined by 5% from January through August, sustaining a trend that first emerged in 2007.

It should be noted that the estimates for the first eight months of 2008 are based on trends in 11 leading European markets, which, according to IPK International, account for roughly 65% of total **European outbound trip volume**.

Sun & beach travel, which dominates the outbound holiday market in Europe, staged a recovery in the first eight months of this year (mainly through the peak summer months, of course), following three years of only modest growth. There were also impressive increases for mountain, countryside and snow holidays, but only a modest rise in demand for touring holidays (perhaps reflecting the high cost of motor fuels) and no growth at all in city breaks, **the boomer of the last years.**

The stagnation in demand for city breaks was attributed by some experts in Pisa to a reduction in demand for multiple secondary trips (many British, for example, have become accustomed to taking five or more a year) – in part due to reduced capacity on low-cost carriers. As discussed below, a number have collapsed this year and others reduced routes and frequencies following the rapid rise in the price of oil and, therefore, fuel prices earlier in the year.

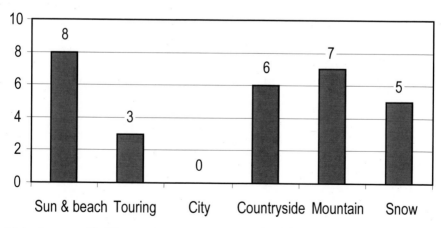

Note: the types of holiday are shown, left to right, in order of their importance

Fig. 9. Trends in European holiday travel by main type of holiday, January through August 2008 (% change on same period in 2007). Source: IPK International's European Travel Monitor

Total European overnight volume on trips abroad, which reached 3.9 billion in 2007, rose by 5% in the first eight months of 2008 – the same as the increase in trips, so the average length of stay remained unchanged at 9.6 nights. Nevertheless, according to the European Travel Monitor, there is estimated to have been an 8% increase in short trips (of 1-3 nights) in the first eight months of the year as against only a 1% increase in long trips (of 4 nights or more) – a trend that is difficult to reconcile with the unchanged overall length of stay, not to mention the stagnation in demand for city breaks.

Total spending on travel rose by 6% in the first eight months of the year, but spending per trip was up 3% and spending per night 2% – not quite keeping pace with inflation. This is a trend which has been evident for the last couple of years. Among the 11 European markets closely monitored by IPK, spending per trip ranged from a high of €1,147 among Spaniards and €1,050 among Russians, to €796 by the Finns and just €724 for Belgians.

5.2   Leading Markets

The slackening demand in recent months seems to have spread across most European source markets, although there are of course big differences in the overall growth ranking in the first eight months of 2008. Russia was the leading market by far in terms of growth, ahead of Finland, with Norway and Spain lagging well behind in third and fourth positions.

According to the European Travel Monitor, Europe's two largest sources of foreign tourists, Germany and the UK (which together account for about 35% of total European outbound trips) achieved very modest growth in the first eight

Table 16. Performance of selected European outbound travel markets, January through August 2008

Market	% change in trips
Russia	15
Finland	9
Norway	4
Spain	4
Italy	3
Sweden	3
Germany	2
UK	2
Denmark	1
Belgium	0
France	−4

Source: IPK International's European Travel Monitor

months of 2008, and the third largest, France, actually shrank by 4%. These trends are broadly confirmed by the arrivals and overnight statistics filed by European Travel Commission (ETC) members on TourMIS, which show very modest increases in arrivals from Germany and France and **even** a slight decline from the UK.

In Germany's case, local experts present in Pisa pointed to the increasing impact of higher airfares, the real estate and financial crises in the USA, and a growing sense of economic unease. The stagnation in the UK market was blamed on a certain saturation in demand for secondary short breaks using low-cost carriers, the gathering crises in the housing and financial markets, the weakening pound and some well-publicised failures of tour operators and airlines. In France, there is a feeling that foreign travel may have been affected by fading confidence, which persuaded many consumers not to commit savings, let alone take on more debt.

It is also clear that destinations in the **Eurozone** suffered from the strong euro, which diverted some of those Europeans who did decide to travel to cheaper non-European destinations. More detailed information on the performance of individual European destinations will be provided by ETC in its annual *European Tourism Insights*, which is expected to be published before ITB Berlin 2009.

Table 17. Performance of selected destinations in the European market, January through August 2008 (% change on same period in 2007)

> 10%				
Turkey	Bulgaria	USA	Thailand	Cambodia
5-10%				
Sweden	Iceland	Germany	Austria	Portugal
Malta	Croatia	Slovakia	Latvia	Estonia
Morocco	South Africa	Cuba	Mexico	Indonesia
1-4%				
Denmark	France	Cyprus	Egypt	Tunisia
Canada	Switzerland	Chile	Dominican Rep.	China
Malaysia	India	Finland		
Weak inbound performance (0% and below)				
Norway	UK	Ireland	Netherlands	Belgium
Australia	Spain	Italy	Romania	Greece
Hungary	Czech Republic	Poland	Singapore	Sri Lanka
Maldives				

Source: IPK International's European Travel Monitor

5.3   Leading Destinations for Europeans

The notion that relative exchange rates (in particular, in the first half of 2008, a strong euro, a weak US dollar and a weakening UK pound) had powerful effects on the choice of destination is only partially confirmed by the ranking shown below. No **Eurozone** destination is listed among the fastest-growing destinations, but several (Germany, Austria, Portugal and Malta), and a number of other destinations in Northern and Central Europe and Scandinavia with strong currencies, did better than average, with arrivals growing by more than 5%.

However, the weak US dollar certainly contributed strongly to the USA's continued **inbound** recovery, with the result that it recorded one of the best rates of growth out of Europe of all destinations. Likewise, Turkey, Bulgaria, Thailand and Cambodia are also very price-competitive, which undoubtedly boosted demand for these countries. Japan (in spite of the weak yen for part of the year), China (in spite of, but perhaps mainly because of, the Olympics), Vietnam (stricken by inflation) and Kenya (stricken by war) fell out of the top ranking.

5.4   Transport

So far in 2008, travel by car has grown the fastest, with an increase of 5% from January through August. The long-established trend for air travel to expand faster has – at least for the moment – come to an end. This is due in large part to the increase in airfares (including any fuel surcharges) caused by high international oil prices, but it also appears to be related to a growing awareness of the hassles and delays involved in travelling by air. Another factor was the retrenchment of low-cost

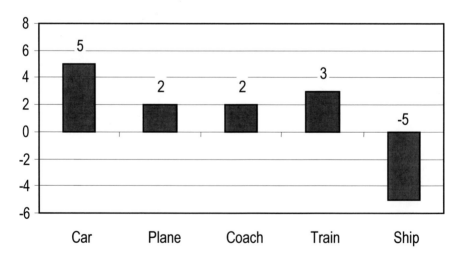

Fig. 10. Trends in European outbound travel by means of transport, January through August 2008 (% change on same period in 2007). Source: IPK International's European Travel Monitor

airlines from routes on which they had expanded over-enthusiastically during the boom.

In fact, for the first time ever, there was a decline in the market share of low-fare flights. Trips using low-fare flights (36% of all trips by air) declined by 1%, while those on so-called 'traditional' airlines (64% of total flights) increased by 3%. However, as IPK points out, it **is becoming** increasingly difficult to draw a meaningful distinction between low-fare and traditional flights.

The recent increase in travel by rail on European outbound trips continued in the first eight months of 2008. This was attributed to the combination of the increased hassles and costs of air travel, the expansion of high-speed rail services in Europe, and perhaps a certain reawakening to the attractions of travelling by train. On the other hand, trips by ship and ferry declined by 5%.

5.5   Travel Distribution

Europeans who use the internet to help with their travel arrangements now outnumber those who do not. Indeed, the share of internet users increased from 50% in January-August 2007 to 55% in January-August 2008. And the use of the internet for online booking continues to rise much faster than its use for simply 'looking' – gathering information prior to booking a trip. The share of online bookings (for at least part of a trip) has risen from 19% of total trips abroad in 2003 to 41% this year.

Table 18. European online travel trends, January through August 2007-08 (% of trips)

	Jan-Aug 2007	Jan-Aug 2008
Online booking	36	41
Online 'looking'	13	14
All internet users	50	55
Non-internet users	50	45

Source: IPK International's European Travel Monitor

Nearly half of all European travellers now use the internet for information about their trips, while the importance of the travel trade as a source of information continues to slide. But the most notable development in this year's responses to this question in the European Travel Monitor questionnaire is a four percentage point increase in those saying they pay attention to friends and relatives – to 19%.

The growing ease of direct bookings through the internet is working to undermine the role of the retail travel trade, but at the same time the rise in dynamic packaging offered directly by tour operators to clients booking online is working in the opposite direction. While nearly 40% of all bookings are made at least partly online, 25% involve travel agents, 10% are booked direct with hotels and 7% direct with

Table 19. Information sources used by European outbound travellers, January through August 2008

Source	% share[a]
Internet	47
Travel agency	23
Friends/relatives	19
Travel guide	8
Travel brochure	7
Newspaper	2
Tourist office	1
TV	1
Others	3
No information	17

[a] Multiple responses possible

Source: IPK International's European Travel Monitor

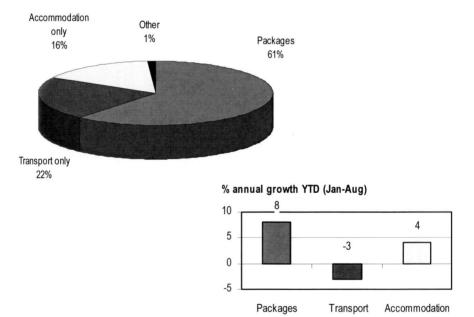

Fig. 11. Breakdown of European pre-bookings, January through August 2008. Source: IPK International's European Travel Monitor

transport companies. Just over a quarter of all European outbound trips do not involve any advance bookings at all.

The share of trips involving full packages continued to rise in the first eight months of this year – by 8%, compared with a 4% increase in the number of 'accommodation only' bookings and a 3% decline in 'transport only' bookings. These full packages include packages self-tailored online, primarily as dynamic packages. Interestingly, although accommodation-only pre-bookings rose only moderately, pre-bookings direct with hotels rose sharply, implying that travel agents and other intermediaries lost some market share in hotel bookings.

Among the full packages, there was a 5% increase in accommodation plus flight bookings, a 7% increase in accommodation plus coach bookings, a remarkable 20% increase in accommodation plus train bookings, and an equally remarkable 16% increase in accommodation plus other bookings. The 'others' might include travel insurance, car rentals, ferries, cruises and excursions.

Among transport-only pre-bookings, coach bookings rose by 6% and train bookings by 3%, but flight and 'other' (presumably mainly ferry and cruise) bookings were both down 4%.

Between Past, Present and Future – Implications of Socio-demographic Changes in Tourism

Philipp E. Boksberger, Kristian J. Sund, and Markus R. Schuckert

Abstract

This paper discusses the possibilities and limits of today's tourism industry analyses regarding the predicted future travel behaviour on the basis of socio-demographic changes. Based on a written survey of German-speaking visitors of a destination in Switzerland, the results support the thesis of cohort-specific travel behaviour. The highlighted changes shall serve as a source for the development of a more diversified supply structure in tourism directed to the mature customer.

1 Between Past, Present and Future

In the year 2050, every fifth person in the world, according to estimates of the United Nations , will be over 60 years old and every sixth person will be over 65 years old (UN 2001). Responsible for this development is the double demographic ageing process (Boksberger & Laesser 2008). On the one hand, the percentage of older people increases because the baby boom generation (the post-WWII high-birthrate generation) had fewer children than their parents. On the other hand, the number of older people increases because of higher life expectancy. The great challenge in tourism lies in the awareness of these demographic changes, their implications for the travel behaviour and their lasting effects (Reece 2004, Nickerson 2000). In order to be able to to quantify and qualify this development, neither assessments of the purchasing power and travel intensity nor traditional segmentation such as "65", "senior citizens/elderly", or "golden agers" are sufficient to cover this trend. In fact, due to the rapid socio-economic developments, every generation now makes its own age and travel experiences under different societal circumstances. With the active participation in the everyday life, senior citizens have not only increasingly become travel enthusiasts but also more discerning and sophisticated in their demands toward the travel experience (Javalgi et al. 1992, Chesworth 2006, Hornemann et al. 2002). There are also tendencies that the demographic ageing of the population is offset by the increased socio-cultural youthfulness of senior citizens (Höpflinger 2005). Many elderly persons nowadays demonstrate a more "youthful" behaviour than previous generations. This trend is closely linked to the developments in economic wealth and the improved socio-political protection of superannuation and comparable schemes. Consequently, tourism research is called upon to generate reliable knowledge and data about the implications of these socio-demographic changes on tourism.

According to the general principles of market research, the approach is clear: If we want to assess how consumption changes when society ages, the group of over 50-year-olds has to undergo closer scrutiny. Do the same principles also apply to tourism or do the conclusions reached today about 50-year-old tourists only reveal very little about the shape tourism will take for mature travellers in the future? Basically, tourism research distinguishes between cohort-specific travel behaviour and age-specific travel behaviour (Lohmann 2007). In the former case, seniors travel in a similar manner to the middle-aged (this is supported by the social learning theory, see Bandura 1977) which would allow for a prognosis based on current travel analyses of 50-year-olds. In the latter case, the travel behaviour of older and younger people is different (this is supported by the lifecycle theory, see Rapoport & Rapoport 1978) which would also allow for a prognosis based on current travel analyses of 70-year-olds.

In order too test these two hypotheses of tourism market research, this paper presents the results of a relevant study and thus supports the discussion with empirical data.

2   Method

The data was collected in 2006, in the German-speaking part of Switzerland, from people whose details are stored on a database of a tourist destination. The survey was conducted using a written questionnaire and mainly comprises questions about individual travel behaviour. Following a promising trial with students, the questionnaire was designed to question people about their current travel behaviour as well as about their travel behaviour ten years ago and their predicted future travel behaviour in ten years time. To ensure the results of the survey were representative, the questions were oriented towards the Swiss travel and tourism market (Bieger & Laesser 2005). The questions also allow for multiple entries that may be ordered according to priority. With a response rate of 37.2 %, an overall sample size of N=701 was achieved. From the overall sample size, persons between 50 and 60 years old (n=141) and persons between 70 and 80 years old were selected. The sample shows the following socio-demographic characteristics and subjective assessments of their own health status by the respondents.

The analysis conducted was based on the frequency of the entries and is listed in the following tables. To test the two hypotheses, the percentage changes for the 50 and 70 year age group with reference to the cohort-specific travel behaviour

Table 1. Socio-demographic characteristics of respondents

		50-60 n=141	70-80 n=45
Sex	Female	46.1%	20.0%
	Male	53.9%	80.0%
Marital Status	Single	15.6%	11.1%
	Married/Life Partner	76.6%	84.4%
	Widowed	3.5%	6.7%
	Family with Children	27.7%	0.0%
	Family with Grandchildren	0.7%	0.0%
Gross Annual Household Income in Swiss Francs	< 70,000	19.1%	40.0%
	70,000 - 90,000	24.1%	31.1%
	90,000 - 130,000	33.3%	15.6%
	> 130,000	22.7%	6.7%
Health Status	Be Fit and Healthy	74.5%	62.2%
	Have Chronic Health Problems	12.1%	6.7%
	Be Overweight (BMI >25)	14.9%	11.1%
	Have Asthma or other Allergy	7.1%	4.4%
	Be Diabetic	0.7%	4.4%
	Have Cardiac and Circulatory Troubles	5.0%	15.6%
	Be Walking Impaired	1.4%	0.0%
	Be Hearing Impaired	1.4%	6.7%
	Be Visually Impaired	2.1%	4.4%
	Be under great Stress / Feel Depressed	6.4%	0.0%
	Need regular Medication	14.2%	46.7%
	Need regular Medical Treatments	7.1%	15.6%
	Suffer any Medical Condition	3.5%	8.9%

"ten years ago", "now/today" and "in ten years" have been itemised and added up in the respective columns. For the calculation of the age-specific travel behaviour, the percentage change of the 70-year-olds "ten years ago" was added to the 50-year-olds "in ten years" according to the same method.

3   Results

The analysis of the motives for travelling shows that "enjoying scenery and nature", "seeking rest and relaxation" and "having time for family, partner, oneself" are the most important reasons in both age groups, as well as "exercising/ sports" and "doing something for your health" that show an opposing development.

Considering the differences in percentage of the absolute value of entries, the analysis reveals a difference between the cohort-specific travel behaviour of 16.56% and the age-specific travel behaviour of 27.50%. From this, it may be concluded that the structure of reasons for travelling consolidates with age and tends not to change later in life. The smallest percentage changes (added difference) in the sample examined, the reasons for travelling are the most stable parameters of travel behaviour.

Table 2. Travel motivation

Travel Motivation	Ten Years Ago		Now/Today		In Ten Years	
	50-60	70-80	50-60	70-80	50-60	70-80
Exercising/Sports	13.7%	11.3%	11.2%	9.4%	8.9%	8.0%
Enjoying Scenery and Nature	15.9%	23.2%	18.2%	23.5%	17.4%	23.2%
Seeking Rest and Relaxation	13.6%	11.9%	14.9%	12.9%	14.9%	15.9%
Sightseeing	5.5%	7.2%	4.8%	4.7%	4.8%	3.6%
Sunbathing and Swimming	9.0%	3.0%	5.9%	1.8%	5.3%	0.7%
Enjoying Gastronomy	6.2%	5.9%	7.0%	8.8%	6.8%	10.1%
Experience History and Culture	7.5%	8.3%	8.8%	10.0%	9.7%	9.4%
Health (Fitness, Wellness, etc.)	3.1%	4.2%	5.7%	5.9%	8.9%	8.7%
Interacting with People/Socialising	1.8%	3.0%	2.9%	3.5%	3.1%	2.9%
Shopping	1.3%	0.0%	1.0%	0.0%	0.2%	0.0%
Having Time for Family, Partner, Oneself	14.3%	10.1%	12.0%	11.2%	11.2%	13.0%
Learning/Broaden One's Horizon	7.3%	10.1%	7.3%	7.1%	7.9%	3.6%
Others	0.7%	1.8%	0.3%	1.2%	0.8%	0.7%

Regarding the question of the preferred type of travel, "holidays in the country-side/ in the mountains" and "winter holidays in the snow" are central to the respondents. "Lake or beach holidays" have experienced the most marked decline. The analysis of type of travel (18.35% vs. 24.54%) also supports the hypothesis of the cohort-specific travel behaviour.

Table 3. Type of trip

Type of Trip	Ten Years Ago		Now/Today		In Ten Years	
	50-60	70-80	50-60	70-80	50-60	70-80
Lake or Beach Holiday	17.6%	9.4%	12.3%	4.9%	12.2%	4.3%
City Break	13.9%	13.0%	14.1%	13.2%	14.4%	9.7%
Excursion by Car/Bus/Coach/Rail	7.9%	10.8%	9.3%	12.4%	10.0%	12.9%
Trip on Cruise/Boat/Yacht	1.9%	5.1%	2.4%	4.1%	3.9%	3.2%
Holdiays in the Countryside/in the Mountains	17.8%	21.7%	19.3%	23.9%	18.1%	27.9%
Spa and Wellness Holidays	2.6%	1.4%	5.6%	4.9%	9.2%	7.5%
Winter Holiday (Snow)	22.5%	21.0%	19.5%	18.2%	15.9%	16.1%
Winter Holiday (Sun and Beach)	3.7%	3.6%	2.6%	5.8%	3.7%	4.3%
Event Trip (Concert, Sport, etc.)	2.8%	3.6%	4.4%	3.3%	3.4%	3.2%
Visiting Friends and Relatives	5.6%	5.8%	7.0%	4.1%	5.9%	5.4%
Special Family Occasion (Wedding, etc.)	2.6%	1.4%	1.6%	1.6%	1.6%	2.1%
Others	1.3%	2.9%	2.0%	3.3%	1.6%	3.2%

Consequently, the same approach was used to examine the details of the booking process, starting with the favourite travel destination, the choice of transport and hotel category and finally the booking of the trip itself.

The analysis of the favourite travel destination illustrates that Switzerland was the most popular travel destination of the respondents ten years ago, today and also prospectively in ten years time. The dominating position of this travel destination results in a marked difference (19.75% vs. 30.10%) in favour of the cohort-specific travel behaviour.

Whereas the sample showed a general tendency towards an increase in travel by train to the travel destination, the analysis highlighted a marked decline of 62.2% to

Table 4. Choice of destination

Choice of Destination	Ten Years Ago		Now/Today		In Ten Years	
	50-60	70-80	50-60	70-80	50-60	70-80
Switzerland	64.5%	75.6%	75.2%	88.9%	68.1%	86.7%
Neighbouring Countries	21.3%	1.3%	14.9%	2.2%	9.9%	2.2%
Rest of Western Europe	2.8%	2.2%	4.3%	4.4%	3.5%	2.2%
Eastern Europe	2.8%	6.7%	2.1%	2.2%	0.7%	2.2%
Africa	1.4%	2.2%	1.4%	2.2%	3.5%	2.2%
USA/ Canada	1.4%	4.4%	4.3%	2.2%	3.5%	2.2%
Central and South America	1.4%	2.2%	0.7%	2.2%	1.4%	0.0%
Middle and Far East	2.1%	2.2%	0.7%	2.2%	1.4%	2.2%
Asia	2.8%	2.2%	3.5%	2.2%	2.1%	2.2%
Australia/New Zealand	0.7%	0.0%	0.7%	0.0%	2.8%	2.2%

Table 5. Means of transport

Means of Transport	Ten Years Ago		Now/Today		In Ten Years	
	50-60	70-80	50-60	70-80	50-60	70-80
Private Vehicle/Rental Car/Mobile Home	66.7%	62.2%	65.2%	51.1%	51.8%	26.7%
Train	19.9%	17.8%	27.7%	40.0%	31.9%	57.8%
Airplane (scheduled or charter flight)	10.6%	13.3%	7.1%	8.9%	9.9%	2.2%
Bus/Coach	6.4%	4.4%	7.1%	4.4%	9.2%	2.2%
Motor Bike/Scooter	1.4%	0.0%	3.5%	0.0%	2.1%	0.0%
Others	1.4%	2.2%	1.4%	2.2%	0.7%	2.2%

Table 6. Type of accommodation

Type of Accommodation	Ten Years Ago		Now/Today		In Ten Years	
	50-60	70-80	50-60	70-80	50-60	70-80
Hotel *+**	5.1%	2.0%	2.2%	2.1%	1.9%	2.5%
Hotel ***	19.7%	24.8%	22.9%	26.1%	20.9%	25.0%
Hotel ****	13.2%	17.8%	21.6%	25.0%	24.1%	26.3%
Hotel *****	3.2%	4.0%	3.8%	4.2%	4.7%	3.8%
Motel/Guest House/Inn/B&B	5.1%	12.9%	6.3%	12.5%	8.2%	10.0%
Own Holiday House/Apartement	7.7%	5.9%	6.3%	3.1%	7.9%	3.8%
RentalHoliday House/Apartement	27.1%	16.8%	23.8%	16.7%	20.9%	16.3%
Camping (incl. Tent, Caravan)	9.4%	3.0%	3.4%	1.0%	3.2%	1.2%
Youth Hostel	3.2%	4.9%	4.1%	3.1%	2.6%	3.7%
Private Accommodation (VFR)	6.1%	7.9%	5.7%	6.2%	5.4%	7.5%

26.7% in the use of private vehicles as the preferred means of transport for 70-year-olds. Although the results of the added changes of 32.25% to 36.30% support the hypothesis of cohort-specific travel behaviour, the calculated difference is the smallest in the entire analysis. Consequently, the choice of means of transport may be regarded as a mixture between cohort- and age-specific travel behaviour.

Contrary to the much-discussed polarisation in the accommodation sector, the analysis results of the accommodation categories show a concentration on three- and four-star hotels. Especially four-star hotels are experiencing an increasing popularity among the 50-year-olds. The calculated difference between the cohort-specific travel behaviour (17.40%) and the age-specific travel behaviour (26.79%) confirms the first hypothesis.

Table 7. Travel booking

Travel Booking	Ten Years Ago		Now/Today		In Ten Years	
	50-60	70-80	50-60	70-80	50-60	70-80
Tour Operator	16.1%	22.0%	15.1%	14.1%	14.1%	16.1%
Travel Agency	31.9%	25.0%	20.6%	18.3%	21.6%	19.6%
Local Tourism Organisation	14.8%	5.9%	19.0%	16.9%	17.0%	14.3%
Carrier/Transport Company	4.9%	10.3%	7.9%	9.9%	7.9%	8.9%
Accommodation Provider	27.8%	32.3%	30.9%	35.2%	30.3%	35.7%
Others	4.5%	4.4%	6.3%	5.6%	9.1%	5.4%

In view of the increasing travel intensity of elderly people (Boksberger & Laesser 2008), booking and the question of efficient information and communication have become the focus of attention. The results demonstrate the increasing importance of direct booking of accommodation. There has also been a noticeable decline of 8.95% in bookings through travel agents in the last ten years. This can probably be ascribed to the increasing experience in travelling and global interconnectedness. Therefore, the analysis of the way in which a booking is made supports the cohort-specific travel behaviour (17.43% vs. 31.65%).

4   Implications of Socio-demographic Changes in Tourism

In the last few years, socio-demographic changes, and with it the increase in the number of mature travellers exhibiting altered travel behaviour, have attracted wider interest in the tourism industry. Analogously, the baby-boom generation, especially due to their enthusiasm for travelling and increased competence as travellers, has become a focus of attention in tourism research (Chandler & Costello 2002, Litrell et al. 2004). The implications of the results of this analysis for the tourism industry are discussed below.

In accordance with Danielsson and Lohmann (2003), it can be assumed that travel behaviour ceases to change or changes only very slowly or insignificantly after reaching a certain age. The findings of this study have confirmed the hypothesis of the cohort-specific travel behaviour. A gradual tendency towards higher levels of comfort and service quality has also been observed (Sund & Boksberger 2008). Only when choosing a means of transport, age-specific behaviour has been observed. The stress-free and comfortable utilization of a transport system is therefore more important than mere cost and time effectiveness. Especially when choosing a means of transport, mature travellers place high demands on quality and safety. A systematic focus on the maturer population in tourism requires an appropriate closed mobility chain (accessibility and disability-friendly facilities on transport systems, accommodation facilities, barrier-free access to other tourist facilities and services) and closed information chains (visual and acoustic orientation aids, service and information facilities).

In general, it has to be stressed that knowing customer demands is essential for planning a target group aligned service and product range for the future. Even regarding the segmentation of the group of mature travellers, the tourism sector is still in need of more detailed insights (Boksberger & Laesser 2008). Undoubtedly, mature travellers should not be viewed as one homogenous group, but should rather be divided according to different interests or activities, such as well-being, art and history, gastronomy, or according to their health (Shoemaker 1989, Horneman et al. 2002, Fleischer & Pizam 2002).

References

Bandura, A. (1977). Social Learning Theory. Englewood Cliffs: Prentice Hall.

Bieger, T., Laesser C. (2005). Reisemarkt Schweiz 2004. Universität St. Gallen: St. Gallen

Boksberger, P., Laesser, C. (2008). Segmenting the Senior Travel Market by Means of travel Motivation – Insights from a Mature Market – Switzerland. Proceedings of CAUTHE 2008, February 11-14, Gold Coast, Australia.

Chandler, J. A.; Costello C. A. (2002). A profile of visitors at heritage tourism destinations in East Tennessee according to Plog's lifestyle and activity level preferences model. Journal of Travel Research 41(2).

Chesworth, N. (2006). The Baby Boom Generation and Leisure and Travel in the Future. In: Weiermair, K., Pechlaner, H., Bieger, T (Eds.). Time Shift, Leisure and Tourism. Impacts of Time allocation on Successful Products and Services. ESV: Berlin.

Danielsson. J., Lohmann, M. (2003). Urlaubsreisen der Senioren. Kiel: F.U.R. Forschungsgemeinschaft Urlaub und Reisen e.V.

Fleischer, A., Pizam, A. (2002). Tourism constraints among Israeli seniors. Annals of Tourism Research, 29(1), 106-123.

Höpflinger, F. (2005). Alternde Gesellschaft – Verjüngte Senioren. Neue Zürcher Zeitung: Zürich [27.09.05]

Horneman, L., Carter, R. W., Wie, S., Roys, H. (2002). Profiling the senior traveler: An Australian perspective. Journal of Travel Research 41(1).

Javalgi, R. G., Thomas, E. G., Rao, S. R. (1992). Consumer behavior in the U.S. pleasure travel marketplace: An analysis of senior and nonsenior travellers. Journal of Travel Research 31(2).

Littrell, M. A., Paige, R. C., Song, K. (2004). Senior travellers: Tourism activities and shopping behaviours. Journal of Vacation Marketing 10(4).

Lohmann, M. (2007). Demographischer Wandel und Konsumentenverhalten im Tourismus – Wie die Veränderung der Altersstruktur die zukünftige touristische Nachfrage beeinflusst. In: Egger, R., Herdin, T. (Hrsg): Tourismus: Herausforderung: Zukunft. Wien, LIT Verlag: 25-44.

Nickerson, N. P. (2000) Travel and recreation outlook 2000: Focusing on demographics. Montana Business Quarterly 38(1).

Rapoport, R., Rapoport, R.N. (1978). Leisure and the Family Life Cycle. London: Routledge.

Reece, W. S. (2004). Are Senior Leisure Travelers Different? Journal of Travel Research 43(1).

Shoemaker, S. (1989). Segmentation of the senior pleasure travel market. Journal of Travel Research, 27(3), 14-21.

Sund, K., Boksberger, P. (2008). Senior and Non-Senior Traveller Behaviour: Some Exploratory Evidence from the Holiday Rental Sector in Switzerland. Tourism Review, in press.

UN (2001). World population prospects: the 2000 revision, New York, United Nations.

Climate Change and Its Impacts on the Travel and Tourism Industry

Tourism 2030: Climate Change Is Re-charting the Map of World Tourism

Eric Heymann and Philipp Ehmer

1 Tourism Industry Is Expanding in Turbulent Times

The international tourism industry has had to face many challenges in the recent past. These include the terrorist attacks of 11 September 2001 as well as those in tourist destinations such as Bali and Egypt. In addition, tourism has been put under pressure by the lung disease SARS, the war in the Middle East and years of rising energy prices. The industry has also been marked by changes on both the supply and demand sides. The travel behaviour of many consumers has changed considerably. Some of the key factors characterising this change are: late bookings, increased price-consciousness, shorter holiday trips, the desire for more flexibility and individuality and the trend towards special and theme holidays. On the supply side, notable changes include the major success of the low-cost carriers and new distribution channels such as the Internet. All things considered, the tourism industry is looking back at turbulent times.

1.1 Average Growth of 4% P.a. in the Sector Since 2000

In the face of these difficult conditions, it is astounding that the tourism industry has been able to achieve extremely high growth during the last few years. For instance, according to the World Tourism Organisation (UNWTO), a United Nations agency, the number of international tourist arrivals increased by an annual average of about 4%, to almost 900 million, between 2000 and 2007. This shows that, in the period stated, the dominant drivers of growth at the end of the day were the dynamic global economy, backlog demand in developing and newly industrialised countries and mankind's inherent desire for individual mobility.

1.2 The New Challenge of Climate Change

In a period of climate change, the tourism industry is now confronted by a new challenge. Unlike natural disasters or terrorist attacks, this is not just a short-term effect that could then be quickly forgotten. Rather, climate change will permanently alter the attraction of some holiday regions and force them to take steps to adapt in the next few decades. It is taken for granted that there will be regional and seasonal shifts in both national and international tourist flows during the next few years. As a result it is also evident: there will be winners and losers from climate change. The remainder of the tourist value creation chain (e.g. tour operators, travel agencies, airlines, hotels) will not be left untouched by this.

The focus of our investigation is the evaluation of particular holiday destinations. To do so, we take into account four factors that influence each of the tourist regions:

- the consequences of the climatic changes, including substitution effects;

- the consequences of regulatory measures to slow climate change and/or to mitigate its negative effects (in particular the increase in the price of mobility);

- the possibilities for adaptation to the changing conditions open to the individual regions;

- the economic dependence of the tourist destinations on (climate-sensitive) tourism.

To start off, we outline the ways in which the environmental-climatic and regulatory-market economy dimensions of climate change can affect the tourism industry. In the final section, using a scoring model, we differentiate between the tourist regions that can profit from climate change and those that are expected to be on the losing side. The forecast horizon of our investigation is 2030.

1.3   Other Factors Are Still Important

Of course, we are aware that not just the factors listed above will be relevant for tourism. The dynamics of the whole economy – in particular the trend and distribution of disposable incomes – as well as external shocks will continue to affect the sector decisively in the future. From a global view, tourism will definitely continue to be a growth sector, due to the pent-up demand already mentioned, rising global incomes and the trend towards increasing freedom to travel (e.g. in China). Up to 2020, we expect an annual average increase of around 3.5 to 4% in international tourist arrivals. Climate change will not lead, therefore, to a shrinking of the tourism industry. This is all the more valid as many types of travel (business travel) and culturally-motivated tourism may continue to be only slightly affected by climatic changes.

2   Climatic Effects on Holiday Regions

Almost all scientists concur that human activities are playing a decisive role in causing and accelerating climate change. According to the Intergovernmental Panel on Climate Change (IPCC), the probable effects of climate change include a rise in the average global temperature, an increase in extreme weather events (e.g. more frequent droughts and heat waves, more storms and heavy rain) as well as a change in regional and seasonal precipitation patterns. For instance, summers in central Europe may become drier on average (nevertheless with increased probability of short-term heavy rainfall), while damper winter months are expected. Precipitation in winter is likely to fall more frequently as rain and more rarely as snow. A further example: the Asian monsoon may strengthen, while the dryness in the remaining seasons will worsen. These phenomena will already be observable before 2030, although they will become more marked in the following decades. In the longer term, an appreciable rise in sea level is expected. Even in the short term, increased damage from storm surges is probable for many of the earth's coastal regions (e.g. as a result of flooding or coastal erosion).

2.1   Differing Ways Climate Change Can Impact – Learning Effect Important

There are a variety of ways in which the environmental-climatic dimension of climate change can affect the attractiveness for tourists and the economic prospects of individual tourist regions. For many holidaymakers – particularly from central and northern Europe – the chance of having "good weather" is one of the most important motives behind the choice of a holiday destination. If in the future the climate – i.e. the "statistical weather" – changes, tourists will learn from their own negative and positive experiences but also from media reports. We consider it very likely that tourists will integrate the changes into their calculations and that they will adapt their travel behaviour accordingly. In the end this will lead to the seasonal and regional shifts in tourist flows already mentioned.

2.2   Examples of Changes in Tourist Flows

The Mediterranean region, with its focus on seaside and beach holidays, loses attractiveness if there is an increased number of heatwaves in the summer months: during the past few years such events have already begun to increase in frequency in the region. People who repeatedly find that their holiday activities are restricted by extreme heat could be inclined to spend future holidays in other regions, or to go to the Mediterranean region in spring or autumn. In contrast, the North Sea and Baltic regions, the northern Atlantic coast of Spain and the Canary Islands are some of the holiday destinations that could become more popular with tourists due to (actual or expected) excessively high temperatures in the Mediterranean region.

Also in winter, shifts in tourist flows are likely. Anyone who frequently experiences lack of snow in the lower-lying ski resorts of the Alps, or the German Mittelgebirge mountains, would tend to switch to higher-altitude or glacier skiing areas in the future. The transfer will be boosted because satisfactory artificial snow creation is not possible in lower-lying regions, or on south-facing pistes, if temperatures are too high. Also, in the future the winter season will be shorter. Of course, a slump in the demand for ski holidays is not expected, so that the higher-altitude ski resorts will increase their market share. Reliability of snow conditions will therefore become more important for the attractiveness of ski areas. By 2030, it is expected that the snow line in the Alps will rise by 300 m. The height above which ski areas can be regarded as having reliable snow conditions will then be around 1,500 metres.

2.3   Damage to Tourist Infrastructure and Attractions

Another way in which climate change affects holiday regions is more frequent damage to tourist infrastructure or particularly attractive regional draws. These could be caused as much by temporary extremes of weather as by the consequences of gradual climate change. More frequent storms and floods, for example, affect facilities like hotels and guest houses.

Although such extreme experiences cannot be exactly forecast, they could nevertheless have an influence on the choice of a holiday destination if there is a pronounced season for extreme weather events (e.g. the hurricane season from June to November in the western Atlantic). Violent storms will accelerate beach and coastal erosion, which must be combated by expensive coastal defence measures to reinforce sections of the coast.

Longer heatwaves, and dry periods that can cause or aggravate natural disasters, are other negative factors for the attractiveness of tourist regions. For instance, large-scale forest fires scare off tourists, or make a visit impossible because of closures: this results in a considerable shortfall in receipts for the period of the fire.

2.4   Other Problem Areas

Already, longer dry periods are causing difficulties for water supplies in some tourist regions (e.g. southern Spain, North Africa, Cyprus) particularly as many tourist facilities (swimming pools, golf courses) and the sheer number of tourists lead to a vastly increasing demand for water. This is in addition to competition for water from agriculture. In many regions, lower precipitation could mean that ensuring an adequate supply of water will be even more difficult or will involve considerably increased costs (e.g. desalination, dams).

Climate change is leading to a warming of the world's oceans. As a result, regions in which diving plays an important part in tourism (e.g. the Red Sea, the

Great Barrier Reef, the Maldives) will lose attractiveness as a result of the bleaching and death of the coral. In the long term – probably well after 2030 – without countermeasures, the rising sea level will endanger the existence of many island nations and atolls in the South Sea and the Indian Ocean (e.g. the Maldives) as well as low-lying coastal areas and cities.

In the future, climate change could also cause more damage to the infrastructure of winter sports regions. As many facilities (e.g. ski lifts) are anchored in permafrost soil, their stability could be endangered if the soil thaws. Increased investment to guarantee safety will probably be necessary in the future.

Lastly, climate change could make preventive measures necessary in the affected tourist areas (e.g. investment in safeguarding the water supply, improvements in coastal protection, more efficient fighting of forest fires) if these regions want to continue to use tourism as a driver of growth and employment.

2.5   Types of Travel Affected to Different Extent

Naturally, different travel types are affected to varying extents by climate change. While the classic summer package holiday in the Mediterranean tourist centres will noticeably suffer from rising temperatures, city trips, which are mainly enjoyed in the spring and autumn months, are generally independent of climatic changes. This also applies to cultural tourism, "wellness" holidays and many other types of theme travel. One thing is fundamentally valid: the more the main reason for selecting a holiday destination is that the holidaymaker hopes for "good weather" or favourable conditions for particular weather-dependent activities (e.g. skiing, diving), the more impact – in both the positive and negative senses – the climate will have on the region concerned in the future. Holiday resorts that will end up more strongly under pressure are those with distinct reliance on only a single (weather-dependent) high season, as possibilities to adapt are then extremely limited. For example, it is hard to attract families with school-age children outside the summer holiday period.

3   Government Measures and Higher Energy Prices Hit the Tourism Industry

According to the IPCC, the transport sector, with its roughly 13% of global greenhouse gas emissions, is contributing considerably to anthropogenic climate change. Of particular importance is the fact that the transport sector has grown rapidly worldwide in the last few years: the notable improvements in the specific energy consumption of the various means of transport have therefore been outweighed by the increased demand. The bottom line is that the proportion of global greenhouse gas emissions from transport is rising. The transport sector is therefore coming under political focus. As the tourism industry is closely interlocked with the transport sector, this industry is also coming under pressure from regulatory measures. UNWTO estimates that the global tourism industry is responsible for about 5% of human-induced climate change.

3.1 Motor Vehicles and Aircraft Are the Most Important Means of Transport

By a large margin, the most popular modes of transport in international tourism are by road and air. Road transport has been in the sights of environmental policy for a long period. The increases in mineral oil taxes in many EU countries during the last few years have also been ecologically motivated. By 2030, a noticeable increase in the rates of mineral oil taxes is expected, especially in Eastern Europe, as the EU strives for a gradual harmonisation of tax rates. However, even in Western Europe, further increases, or higher charges for toll roads, are likely. This is on top of the increase in crude oil prices expected in the next few years.

It is true that the energy efficiency of cars will steadily increase in the next few years – dependent also on EU directives for the reduction in CO_2 emissions by new cars. However, in the long term the price effects on petrol will have a stronger effect than savings following technical progress in vehicle and engine design. The longevity of cars also means that a sudden adjustment to increasing fuel prices is technically scarcely possible. All in all, therefore, by 2030 motoring will be more expensive.

3.2 Including Air Transport in Emissions Trading

Concerning journeys of 1,000 to 1,500 kilometres or more one way, the choice of transport is likely to be by air rather than road. The liberalisation of European air transport, which paved the way for the enormous success of the low-cost carriers, has meant that many European destinations are now accessible and affordable even for private households on relatively low incomes.

However, air travel is currently being more intensively scrutinised by environmental policy. The European Commission plans to bring the sector into EU emissions trading by 2012 at the latest. The message is clear: the firms will be faced with costs that, depending on the way emissions trading is organised (e.g. scarceness of certificates, allocation mechanism), will lead to higher ticket prices. In air transport as well, technical advances will not be able to react sufficiently to the rising prices: aircraft have even longer operational lives than cars. However, the introduction of the new, extremely efficient generation of aircraft (B787, A380 and A350), the (long overdue) realisation of a "Single European Sky" and the further liberalisation of global air transport could slow the trend towards increasing ticket prices.

The bottom line is that flying will be more expensive in the future, as a result of regulatory measures and increasing fuel prices. The additional costs could influence the choice of a holiday destination, particularly for families taking intercontinental flights. Ceteris paribus, long-haul flights will in any case be more heavily affected than e.g. air travel within Europe.

3.3   Rail and Sea Travel Among the Winners

Railways and ships are considered to be environmentally friendly modes of transport. In the current political environment, no regulatory measures to burden these modes of transport are being planned. Significant implications for the price of cruise tickets are not anticipated: in addition, the clientele for traditional cruises is usually affluent. For rail travel, environmentally-motivated reliefs are even conceivable in the next few years. However, rail companies will also have to contend with increasing energy and fuel prices. Overall, the relative price of rail travel could fall, in comparison with road and air travel. The expected further opening of the market in European rail transport and the increase in intensity of competition could, as with air travel, have a slowing effect on the trend of prices. In the longer term this could encourage innovative products, particularly in – currently still insignificant – cross-border passenger transport.

This means that regions that can be accessed by rail could benefit somewhat in the future. If such holiday areas also have attractive tourist products for cyclists, there is growth potential in the niche market. Nevertheless, up to now rail transport has only played a secondary part in holiday travel.

3.4   Long-Haul Destinations Will Be Put at a Disadvantage

The increasing price of mobility will affect longer-haul holiday destinations more than closer ones. As the industrial countries in the heart of Europe (in particular Germany, the U.K., France), the USA and Japan are some of the freest-spending nations, regions that are well away from these source markets are likely to be the first to lose. Climate-policy motivated government measures will reduce the purely economic attraction of long-haul destinations. In contrast, short-haul destinations will be relatively favoured. In addition to the increasing price of mobility, higher energy prices for tourist facilities (hotels, swimming pools, snow cannons, air conditioning, leisure parks etc.) will also play a part.

4   Assessment of Individual Tourist Regions

In the following, we will take a more detailed look at individual countries and tourist regions, as well as their susceptibility to climatic changes. We focus on Europe as it is the world's biggest travel market. Of course, we realise that the degree of differentiation of this analysis is not sufficient in many cases: the regional tourist centres vary too much to allow this. However, it is possible to identify resilient trends for all the countries and regions that have been studied.

Europe is the most important tourist region in the world. According to UNWTO, in 2006 nearly 55% of all international tourist arrivals (461 million) were on the "old continent". In the following comments, we concentrate mainly on the changes in climate and, in passing, on the possibilities for adaptation. The increasing price of mobility is less significant, as distances in Europe are of manageable dimension. After all, nearly all regions will be affected to a similar extent.

4.1   Southern Europe and Mediterranean Regions: Trend to the North

Southern Europe and the **Mediterranean region** are the favourite holiday destinations in Europe. According to UNWTO, in 2006 about 165 million tourists visited these regions. Climatic changes may affect the various Mediterranean states in a similar way. The key factor for the attractiveness of this region is its Mediterranean climate. In the future, increasing average temperatures, together with the increasing probability of heatwaves, could result in temperatures exceeding comfortable levels more frequently. A further problem that many areas may have to confront is a shortage of water, resulting from lack of precipitation and the increasing use of irrigation in agriculture. This restricts the operation of tourist facilities (swimming pools, golf courses). In addition, the increasingly dry summers enlarge the risk of forest fires in many areas.

4.2   The Southern and Eastern Coasts of Spain Are Among the Losers

Spain is – in terms of international tourist arrivals – the second favourite holiday destination after France, with a global market share of approx. 7%. The country has a high proportion of foreign visitors (2006: 59%). The tourism sector, with its very high proportion of GDP – currently about 17% – makes a considerable contribution to Spain's economy.

Spain's most developed tourist areas are close to the Mediterranean. In 2005, a quarter of international tourism was to the Catalonia region. Compared with the city of Barcelona, the Costa Brava and the Costa Dorada attract mainly seaside holidaymakers from northern and central Europe. The second and third most important holiday regions are the Balearic Islands in the Mediterranean (in particular Majorca) and the Canary Islands, off the Atlantic coast of Africa. Next come Andalusia and the Valencia region, which also borders the Mediterranean. According to the Spanish government, these five regions account for more than 80% of international tourist arrivals.

In the future, these tourist destinations will suffer from more frequent heatwaves, which will put off seaside tourists in the important high season. In addition, there could be problems with water supply, particularly as agricultural irrigation is playing a more important part in southern Spain. It is principally the

Spanish mainland that is affected by climate change. According to estimates, temperatures could rise more sharply there than in other countries bordering the Mediterranean. Although Andalusia in particular has many alternative attractions apart from purely seaside holidays (e.g. the Sierra Nevada, Granada, Seville), in the end the success of tourism in the whole Spanish coastal region is based around the beaches.

4.3   The Canaries Could Benefit

In contrast, the effects on the Canary Islands will be less pronounced. Their increased proximity to the equator and subtropical climate mean that temperatures will not rise so much and the differences between the summer and winter seasons will remain relatively small. Even in the future, this will guarantee balanced occupation of tourist capacity over the year and will increase the independence of this holiday destination from climate change.

However, the Canary Islands are suffering from increasing susceptibility to forest fires. On some islands (e.g. Fuerteventura and Lanzarote), water supply could be made even more difficult and expensive (desalination, tanker ships).

4.4   North Atlantic Coast Is Likely to Catch up

The winner in Spain is the northern Atlantic coast. Here, an increase in the moderate temperatures and lower levels of precipitation could have a positive effect on the attractiveness of holiday regions that so far have mainly been favoured by domestic tourists. However, with its present approximately 5% share of Spain's international tourism, and considerably lower tourist capacity in comparison with southern Spain, the Atlantic coast will not be able to compensate for the setbacks in growth or losses of turnover suffered by the Mediterranean region. Overall, Spain's tourism industry will therefore be among the losers from climate change. City holidays to Spain (especially Barcelona and Madrid) will generally be unaffected by climate change, as most visitors already travel there out of the summer month season.

4.5   France's Variety Has a Positive Effect

In terms of international tourist arrivals, **France** is the world's favourite holiday country. According to UNWTO, in 2006 79 million travellers arrived there (9.3% share of the world market in 2006). Only 36% of the tourists in France are foreigners. In France, tourism accounts for around 9% of GDP, roughly in line with the global average.

In France, the Mediterranean region, with Provence and the Cote d'Azur, is particularly well developed for tourism. Even if climate change were to have the expected negative consequences, after taking into account substitution effects the region could benefit. The increase in summer temperatures here could be less serious than in the still hotter country of Spain. Tourist flows could therefore be diverted from these countries to climatically similar – but on average more pleasant – locations, like the south of France.

4.6   Many French Holidays Are Not Dependent on the Climate

A large proportion of tourism in France is largely independent of climate. City holidays to Paris would therefore be as little affected as cultural holidays (e.g. visiting the chateaux of the Loire). The Massif Central and the hinterland of the Provence are likely to be relatively unaffected by climatic changes, at least until 2030.

The French Atlantic coast could benefit from climate change. Higher temperatures and lower levels of precipitation could extend the summer season and make the sometimes harsh climate more pleasant for sea and sand holidays. Apart from that, many tourists are drawn to this region primarily for its variety of landscapes.

4.7   Reliable Snow Cover in the French Alps

Winter sports tourism in the French Alps could be left largely unscathed by climate change for the moment. Many important ski areas are at high altitude: until 2030 lack of snow should normally either be no problem or could be compensated for by artificial snow production. In addition, substitution effects from other ski areas in the European Alps could mean that the winter sports areas in the French Alps will gain. In contrast, the reliable snow cover in the French Pyrenees is noticeably reducing.

Overall, tourism in France could benefit from climate change. The world's favourite holiday country has a sufficiently diversified tourism structure. Besides the "summer sun, sea and sand" theme on the Mediterranean it has other options that are independent of climate, or could even benefit from climate change. The low proportion of international tourists provides a degree of stability, as domestic holidaymakers are usually less flexible in the choice of their destinations.

4.8   Italy Has a Diversified Structure of Tourism

Italy is in third place in Europe and globally in fifth place in the ranking of favourite holiday countries, despite the fact that the country has had to accept considerable downturns in international arrivals in the last few years. Foreigners account for 43% of overnight stays. The tourism industry generates just under 9% of GDP.

Italy also has a strongly diversified structure of tourism. In addition to seaside holidays, e.g. on the Adriatic in Tuscany and on the Italian Riviera, which together account for about a quarter of international arrivals, culturally motivated city tourism constitutes the lion's share, with well over a third (e.g. Rome and the cities in

Tuscany). Other important attractions are the Alps (in particular Alto Adige) and the Italian lakes (especially Lake Garda and Lake Maggiore) in the north of the country.

Italy's revenue from tourism is mainly generated in the north (including the Alps, the coasts and Tuscany) and the centre of the country, including Rome. According to the Italian office of statistics (Istat), southern Italy and the islands notch up barely 20% of tourist arrivals. Foreign tourists have an even stronger preference for the north of the country: this is unlikely to change in the future. This is because rising temperatures e.g. on the southern Amalfi coast and in Sicily could have a more serious effect than in Tuscany and the lakeland regions, where the climate is milder. The fact that international tourism is already concentrated at a higher latitude (comparable with Provence) leads to the assumption that Italy will be less disadvantaged by the effects of climate change. The shifting of tourist flows further to the north within Italy could continue, so that regions already having weak economies must be prepared for more serious setbacks.

The very low altitudes of many ski areas mean that winter sports tourism in the Italian Alps could well be more seriously affected by climate change than those in France. About half of the ski resorts are below 1,300 m. The location on the south side of the Alps means that even the higher-lying areas (e.g. in the Dolomites) are suffering from reduced snow reliability. However, the largest proportion of holidaymakers, about two thirds, visits in the summer season (April to September). In this period, the pleasant temperatures may mean that the Alpine region could even benefit.

The bottom line is that the extensive range of tourist destinations, partly independent of the weather, together with the possibility to attract tourists from regions whose climates will be worse affected, mean that the effects of climate change on Italy should be manageable.

4.9   Central Europe Reaps the Benefits

In comparison with the Mediterranean region, **central Europe** mainly has other types of tourist activities and therefore can also expect differing consequences from climate change. On the one hand, increasing temperatures could make tourist destinations there more attractive. On the other hand, they will lead to precipitation falling more frequently as rain and less often as snow during the winter months, shifting the snow lines to higher altitudes. This could endanger winter sports holidays in many mountain regions.

4.10   North Germany – Europe's New Beach Destination?

The major tourist regions in **Germany** are in the north and south of the country. According to the DRV, in 2006 10 million Germans took their holidays on the North Sea and Baltic coasts: this is about a third of the total. The statistics show that 9.3 million Germans (31%) took holidays in Bavaria and Baden-Württemberg. For the Germans, therefore, their own country is still the favourite holiday destination.

In 2006, measured by international tourist arrivals, Germany was in seventh place globally and in fifth place in Europe. The tourism industry accounts for approx. 8% of German GDP.

No negative effects from climate change are expected either for activity holidays or for seaside holidays on Germany's coasts – on the contrary: there could be positive effects resulting from the longer summer season. The North Sea and Baltic coasts will be favoured by climate change.

In our view, the proportion of overnight stays taken by foreign tourists (currently 15%) will increase, as it should be possible to attract foreign holidaymakers to North Germany from the hot Mediterranean region in the summer months. For German nationals this applies anyway. A positive factor is that city tourism, which currently accounts for approx. 15% and is growing, will remain as unaffected by climate change as will the widespread tourism for treatment at health resorts.

4.11 Winter Sports in Germany on a Knife Edge

However, the German Mittelgebirge mountains will be affected by a lack of snowfall. As early as 2030, many regions might be without snow, or a least may have to contend with a shorter season. The winters of 2006/07 and 2007/08 may have given a foretaste of this. Even in the Alps, only the higher-lying winter sports resorts may be able to escape this general trend. It is highly questionable whether the winter sports regions will be able to compensate for any losses in winter by increased numbers of holidaymakers during the summer months. In winter, it is likely that holidaymakers will prefer alternative resorts, e.g. in Switzerland.

Nevertheless, the bottom line is that climate change is likely to have positive consequences for tourism in Germany. The Potsdam Institute for Climate Research Impact is predicting that climate change will result in long-term growth in demand, of the order of 30%. A positive factor for Germany is the very short distances to holiday resorts, not only for German citizens but also for Scandinavians and residents of the Benelux countries.

4.12 The Altitude Makes the Difference in the Alpine Region

Over the whole **Alpine region**, an increase in average temperature of 1°C could reduce the proportion of ski areas with reliable snow from today's 91% to about 75%. An increase of 2°C could bring this down to just under 61%. A transfer of tourist flows from the lower-altitude resorts to higher-lying holiday regions is very likely. There could be a redirection of tourists, from Germany, Italy and Austria to Switzerland and France, because of their larger number of skiing areas with reliable snow. In summer, the whole Alpine region will increase its attractiveness.

5   Repercussions on the Tourist Value Creation Chain

The climatic changes discussed above and their consequences for holiday destinations have repercussions on the tourist value creation chain. In addition to the regional and seasonal shifts of tourist flows, in the future holidaymakers could wait until later before booking their travel. Travel agents and tour operators must therefore develop stronger incentives for earlier bookings: price could well be the most important factor. The increasing risk of extreme weather could be diminished by the use of specific financial products. Weather derivatives, promising financial compensation in the event of "bad weather", are possible, as is holiday cancellation insurance that would allow the holidaymaker not to set out on the journey if weather conditions at the holiday destination are likely to be poor.

Hotels and tour operators will have to concentrate their own marketing activities on attractions that are less climate- and weather-sensitive. Theme holidays (e.g. "wellness" and health, golf, hiking or cycling, literature, culture, viticulture) are possible alternatives. This would enable tourist flows to be more evenly spread over the year. However, it is also obvious that such diversification would not be successful in every holiday region or for every hotel. Large hotels in the Mediterranean region have a particular problem, as high fixed costs often mean that it does not pay to open the hotel for only a few holidaymakers.

For the air-traffic industry, more uniform seasonal tourist flows would smooth out the normal current peaks of demand in the summer months: this would be beneficial for the sector. Depending on how it is set up, emissions trading could cause disadvantages in terms of price competition for European airlines on intercontinental flights. This would be the case if connecting flights in the destination country were not subject to emissions trading, unlike the direct flights from Europe that certainly would be. However, for the time being the negative effects should still be limited.

6   Conclusion: Winners and Losers from
Climate Change

In order to better assess the countries in which the tourism industry will benefit from climate change and those in which negative effects can be expected, we have compared the most important countries with the aid of a scoring model. This model is based on four parameters, which we have assessed, quantitatively or qualitatively, for all the countries: firstly, direct climatic effects; secondly, substitution effects resulting from climate; thirdly, regulatory burdens and consequent geographical substitution effects; and fourthly the possibilities each country has to adapt to climate effects. The time horizon is 2030. In a subsequent step we have also identified the countries in which economic reliance on climate-relevant tourism is particularly high, as the overall economic effects are particularly relevant for these.

On the losing side in Europe are, in particular, the countries bordering the Mediterranean, with the countries in the eastern Mediterranean being particularly negatively affected. Those that could gain include the Benelux countries, Denmark, Germany and the Baltic States. Scandinavia and Great Britain are also amongst the winners. As a result, the North-South alignment of Europe will be weakened. An interesting factor is that, up to 2030, despite the completely negative effects of higher temperatures in the Mediterranean region, France and Italy will benefit slightly as a result of their diversified tourism structures, as described above. Up to 2030, Switzerland will also be one of the gainers, in particular due to its high snow reliability in comparison with the other Alpine countries, while Austria will tend to lose.

Outside Europe, most countries will suffer from climate change, albeit to differing degrees. Climate change predominantly means additional burdens for all the poorer countries in our investigation that are putting great hopes on tourism as a driver of development. Canada, New Zealand and – with reservations – the USA are the only three further countries outside Europe whose tourism industries will be on the winning side.

At this point we must remind the reader that our investigation is only a partial analysis. We therefore do not expect that the turnover from the tourism will inevitably fall in countries that are on the losing side of our scoring model; in most instances the sector could even continue to expand, for the reasons stated earlier. At the same time, our results can be seen as a warning signal, as it is impossible that the climatic situation after 2030 will improve in the countries that have negative prospects. This is not unimportant, as tourist infrastructure that is being invested in today (e.g. hotels, ski lifts) will mainly still be in existence after 2030. The possibilities of their continued use after 2030 should therefore already be taken into account today when making investment decisions, not least because there are frequently no possibilities for adaptation. Another factor is that many climatic consequences will not take effect until after 2030.

Of course, our scoring model has limits. After all, uncertainties concerning the extent, speed, and specific consequences of climate change are still too great. There are regional peculiarities in individual countries, particularly in the countries outside Europe, that we have been able to reflect only fragmentarily. By focusing on the dominant form of travel in each of the countries under investigation, we have only partly addressed this problem. Also, in many countries there are individual regions that are very differently affected by climate change. The aggregate values from our scoring model at country level are therefore an average, which cannot tell the whole story. Finally, we nonetheless consider that results are reliable on the whole, provided that they are understood as being indicative of trends. In addition, we have recalculated our model using different weightings, with no change to the basic conclusions. The core conclusions are accordingly robust:

Countries with High Economic Dependence on Tourism Are Particularly Affected.

Negative climatic consequences always have particularly serious effects where climate-sensitive tourism has a large economic weighting. In Europe this is in Malta, Cyprus, Spain, Austria and Greece. In the Caribbean, this category includes e.g. the Bahamas and Jamaica. In Asia this applies to Thailand and Malaysia; in Africa to Tunisia and Morocco. Countries particularly dependent on (climate-sensitive) tourism are the island states in the South Pacific and even more so those in the Indian Ocean (primarily the Maldives and the Seychelles). Even though the climatic effects will not be "life threatening" there until 2030, the message is however clear: if climate change means that the tourists stay away, there are considerable negative effects on the whole economy. Among the countries that will experience positive climatic effects by 2030, Estonia, Slovakia, Switzerland and New Zealand are the most dependent on tourism. In the European countries, in Canada and in the USA, tourism is of below average significance. Due to the expected substitution effects, the significance of the tourism industry in many "winning" countries could, however, increase in the coming decades as a consequence of climate change.

This article is based on a study of Deutsche Bank Research. Please download the complete study at: www.dbresearch.com.

Fig. 1. How climate change influences tourism destinations

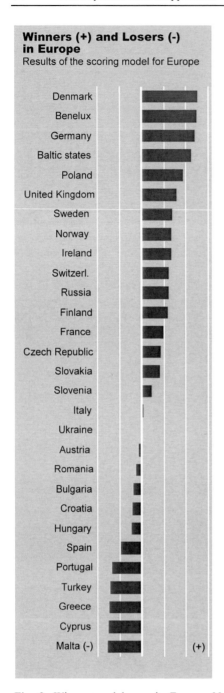

Winners (+) and Losers (-) in Europe
Results of the scoring model for Europe

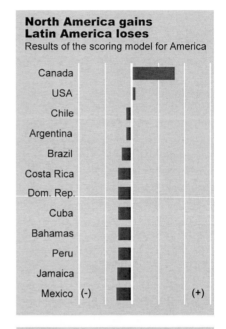

North America gains Latin America loses
Results of the scoring model for America

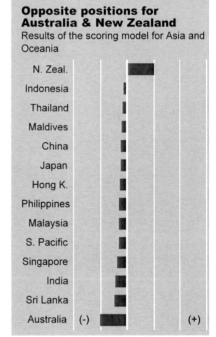

Opposite positions for Australia & New Zealand
Results of the scoring model for Asia and Oceania

Fig. 2. Winners and losers in Europe, North and South America, Asia and Oceania. Source: DB Research

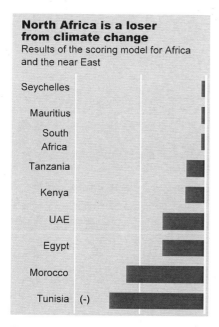

Fig. 3. Winners and losers in Africa and the Near East. Source: DB Research

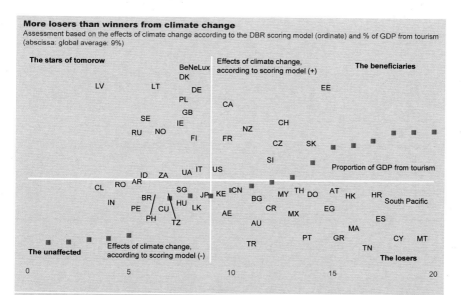

Fig. 4. Countries affected by the climate change. Source: DB Research

2030: Alps Tourism in the Face of Climate Change

Hansruedi Müller

In recent times discussions on climate protection have reached a level never seen before. Tourism has often been a topic of focus in these discussions. Increasingly tourist organisations have realised that they must address climate change, its consequences and especially their course of action as regards tourism. Conferences have been planned, workshops organised, work groups established and publications prepared. The Bernese Oberland and the Switzerland Tourism commissioned the Research Institute for Leisure and Tourism (FIF) of the University of Berne to develop a baseline study together with tourist specialists in order to move the discussion forward. The following is a summary described through seven premises:

Premise 1: Management of Change Is a Central Challenge

Decision makers in the tourism industry face great challenges: Globalisation is intensifying pressure of competition, structural problems are intensifying pressure for change, weakness in profitability is intensifying pressure on financing, technical development is intensifying pressure on innovation, changes in values is intensifying pressure to adapt, and climate change is intensifying pressure to diversify. Trend studies on travel behaviour have repeatedly revealed a similar pattern: Travellers are becoming more individually oriented, more spontaneous, more experienced and more demanding. People travel more often, for shorter periods and as inexpensively as possible. In addition, travel itself should continuously become safer, more exotic, more exciting and at the same time more relaxing. The trend shows that tourists are becoming more environmentally sensitive, however they hardly act environmentally responsibly, and when travelling exhibit a very opportunistic concept of the environment: Environmental damage is only then recognized when their own holiday enjoyment is threatened. And careful: Various environmental phenomena, such as climate change, cannot or can hardly be perceived by the senses.

Premise 2: Global Climate Change Is a Nearly Undisputed Fact – It Is Getting Warmer and in the Winter There Is More Rainfall, in the Summer Less

The Spring 2007 Climate Report of the Intergovernmental Panel on Climate Change (IPCC) forecasts a temperature increase of between 1.8 and 4°C by the end of the 21st Century as compared to 1990, depending on the models and scenario used. In the Alps the temperature will increase even more compared to the global mean temperature. The IPCC estimates that temperatures will increase by up to 2.6°C in summer, and up to 1.8°C in winter already by 2030 (Frei 2004). This increase is due, among other things, to the fact that temperatures over land masses increase more than the global mean, reduction in snow cover in the mountain areas contributes to additional warming, and higher locations and/or latitudes tend to exhibit stronger temperature increases.

Aside from the warming, changes as regards rainfall should also be taken into consideration: While the winter and spring will probably have higher rainfall, the summer and fall will be drier. There is an increasing danger that water will become scarce for making artificial snow in the fall.

Premise 3: Tourism Is an Important Source of CO2 Emissions

Tourism is not only affected by, but also an important co-contributor to climate change. Globally tourism contributes about 5% to CO^2 emissions, which is domi-

Table 1. Temperature change 1990 – 2030 (in degrees °C)

	Seasons	Probability		
		0.025	0.5	0.975
North Switzerland	Winter (Dec.-Feb.)	0.4	1	1.8
	Spring (March-May)	0.4	0.9	1.8
	Summer (June-Aug.)	0.6	1.4	2.6
	Fall (Sept.-Nov.)	0.5	1.1	1.8

Source: Frei 2004

Table 2. Rainfall change 1990 – 2030 (in percent)

	Seasons	Probability		
		0.025	0.5	0.975
North Switzerland	Winter (Dec.-Feb.)	−1	+4	+11
	Spring (March-May)	−6	0	+5
	Summer (June-Aug.)	−3	−9	−18
	Fall (Sept.-Nov.)	0	−3	−8

Source: Frei 2004

nated by road traffic (32%), air traffic (40%) and lodging (21%) dominate. This represents a disproportionate share of emissions when compared to tourism's contribution to the global gross domestic product of 3.6%. (UNWTO 2007)

In addition, tourism is regarded as a global growth sector, which is why in the future its share of CO2 emissions will grow, thereby endangering the Kyoto objectives more greatly than other sectors. Due to the fact that mobility is a requisite for tourism, air traffic will continue to gain in importance. Air traffic has grown tremendously in the past 20 years, and this trend remains despite market turbulences. Currently airlines transport 1.6 billion passengers and 30 million tons of cargo annually. In 1980 fuel consumption was approximately 33 m tons of kerosene, while currently it is 205 m tons per year (producing 300 m tons of greenhouse gases). If this trend continues, it will reach 600 m tons in 20 years. The consequences of this massive increase in air traffic for the environment are extensive. The greenhouse gases emitted in the upper atmospheric layers are three times as environmentally harmful as emissions near to the ground. Proportionate to the environmental damage, aviation fuel (still tax-free) would have to be valued three times as high as petrol or diesel.

Premise 4: Tourism Is a Major Victim of Climate Change – Afflicted by an Absence of Winter Atmosphere, Melting Glaciers, Increasing Natural Hazards

In the 1990s, ski areas under 2000 m a.s.l. experienced by far the least snowfall since 1930. (Laternser/Schneebeli 2003). With more winter rainfall, in the future snowfall will increase in the higher elevations, while in lower elevations precipita-

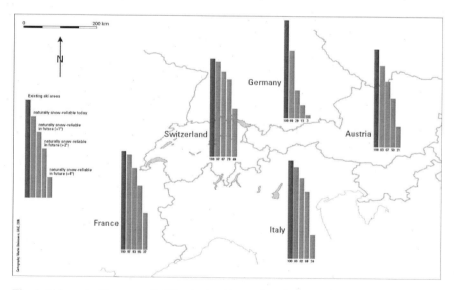

Fig. 1. Ski areas with snow reliability in the Alps region. Source: Abegg, OECD, 2007

tion will more often fall as rain. Thus, even with an overall mild climate more snowfall is to be expected in high elevations. However, in deeper lying regions the winter atmosphere is increasingly absent. Additionally, later heavy snowfall and thus shorter winters are expected.

In international comparisons it appears that Switzerland, even with a stronger warming, will have ski areas with snow reliability for a longer period than in neighbouring countries because these ski areas tend to lie higher. Particularly Germany, but also Austria, have a competitive disadvantage in this regard. With a warming of 2°C, approximately 80% of the ski areas in Switzerland still have adequate snow available, while in France only 65%, in Italy 68%, in Austria 50% and in Germany just 13% of the ski areas will have snow reliability.

Between 1850 and 2000 the volume of all glaciers in Switzerland shrank by approximately 50% (Spreafico/Weingartner 2005). Today there are only about 2000 glaciers remaining, covering approximately 1050 km^2. That makes up 2.5% of the area of Switzerland. During the extremely hot summer of 2003 alone, the alpine glaciers lost an additional 8% of their remaining volume. New studies show that with a warming of 2°C compared to the period 1971-1990, approximately 65% of the alpine glacier area will be lost (Zemp, Haeberli et.al. 2006, S. 3). This will particularly shape the development of the landscape, not to mention the water supply and the type and nature of dangers.

There are many indications that the warming of the atmosphere will also affect the intensity and frequency of weather extremes. Freak weather and extreme events will increase in all likelihood. Although individual extreme events cannot be directly linked to climate change, the frequency of various types of events will still increase. For example, more frequent heat waves, higher risk of flooding, as well as more avalanches and mudslides are to be expected.

When precipitation decreases in the summer and fall, droughts are the result. On the other hand, in the winter half-year precipitation increases, falls more frequently as rain instead of snow and is therefore not bound. Thus, with the snowmelt an increase in the flood frequency is likely in the spring. Water shortages in the Alps regions, though, will hardly pose an acute problem.

Premise 5: Climate Change Also Brings Opportunities

This brief description of possible effects shows that climate change entails some risks for tourism. But it also carries with it opportunities that can be taken advantage of. These can be quite different depending on the region and type of tourism. Possible opportunities and risks for the tourism sector are outlined below (cf. Table 3).

The opportunities and dangers named above are to seen as general tendencies. How strongly these impact tourism destinations greatly depends on local conditions and the degree of adaptation. In addition they are only likely within a certain climatic spectrum. In the event of even stronger climate change, aspects that are seen as opportunities for tourism today could just as well become dangers tomorrow.

Table 3. Climate change: opportunities and dangers

Opportunities	Dangers

• Increase in appeal
  — by offering summer retreats for mountain tourism
  — through seaside tourism in lake regions
  — with more Mediterranean-like conditions in cities
• Improved competitive situation
  — for snow sports locations in higher elevations
  — due to changed climatic conditions in the regions of competitors (e.g. Mediterranean region too warm)
• etc.
	• Loss in appeal
	  — due to the absence of winter atmosphere in the source areas
	  — because of reduced snow reliability in the foothills of the Alps
	  — through changes in the landscape (e.g. through glacial recession)
	• Increasing need for investments
	  — to adapt tourist offerings to changed conditions (e.g. air conditioners)
	  — to ensure snow reliability
	  — to protect against natural hazards (risk management)
	• etc.

Premise 6: Reduction Strategies (Mitigation) Are an Ethical Duty – Everyone Should Contribute in Equal Measure

Tourism is not only an affected party, but rather also a perpetrator. The following core strategies for reducing climate gases can be identified, whereby the prioritisation uses the following logic:

Consume less (fossil) energy, utilise energy more efficiently, switch to renewable energy sources, offset emissions of greenhouse gases. The different reduction strategies can be subdivided into five core strategies:

1. Reduction of energy consumption resp. of the CO^2 emissions

2. Promotion of public transport – optimisation of traffic management

3. Direction via financial incentives – consistent application of the polluter pays principle

4. Compensation of CO^2 emissions

5. Strengthening of communication

Only a combination of all five strategies leads to success.

Premise 7: Adaptation Strategies (Adaptation) for Tourism Are Many-Layered – Climate Change Is a Pressing, but Gradual Issue

Furthermore, tourism must continuously adjust itself to climatic changes and adapt itself to changing conditions in order to remain competitive. The willingness to change is a fundamental requirement for being able to meet new challenges in the dynamic business of tourism. The varied adaptation strategies can be subdivided into three areas, with a total of six core strategies:

- Development of the product range
  - Promotion of innovation and diversification, increase of research
  - Further development and protection of snow sports
- Minimisation of dangers
  - Strengthening of danger defence through technical measures
  - Risk reduction through organizational measures
- Communication
  - Clear positioning and targeted marketing
  - Raising public awareness

To bear in mind: Both climate change as well as the development of tourism will only reach the desired turning point when we do not merely place our hope in someone else's hands, but make a convincing start ourselves. Even the smallest step in the right direction has a value unto itself: A small personal revolution as a kick off and precondition for greater change.

References

Frei, C. (2004): Die Klimazukunft der Schweiz – Eine probabilistische Projektion, Institut für Atmosphäre und Klima, ETH Zürich

IPCC (2007): Climate Change 2007: The Physical Science Basis, Contribution of Working Group I to the Fourth Assessment Report of the Intergovernmental Panel on Climate Change, Cambridge University Press, Cambridge, UK and New York, NY, USA

Laternser, M., Schneebeli, M. (2003): Long-term snow climate trends of the Swiss Alps (1931-99), International Journal of Climatology 23/2003, 733-750

Müller, HR., Weber, F. (2007): Klimaänderung und Tourismus – Szenarien für das Berner Oberland 2030, FIF Universität Bern (eds.), Bern

Müller, HR, Weber, F. (2008) 2030: Der Schweizer Tourismus im Klimawandel, ed. Schweiz Tourismus, Zürich

OECD (2007), Abegg, B., Agrawala S., Crick, F., De Montfalcon, A. (2007): Climate Change impacts and adaptation in winter tourism. In: Climate Change in the European Alps, OECD-Study, Agrawala (publ.) Paris, 25-60

UNWTO, UNEP, WMO (2007): Climate Change and Tourism: Responding to Global Challenges, Advanced Summary, October 2007

Spreafico, M., Weingartner, R. (2005): Hydrologie der Schweiz: Ausgewählte Aspekte und Resultate, Bundesamt für Wasser und Geologie (BWG) (ed.), Bern

Zemp M., Haeberli W., Hoelzle M., Paul F. (2006): Alpine glaciers to disappear within decades? In: Geophysical Research Letters Vol. 33, L13504, Bern July 2006

Zero Footprint – A Viable Concept for Climate-Friendly Tourism in Africa?

Ralph Kadel, Christoph Kessler, and Laura Vogel

1 Introduction

The tourism industry is a strong sector in the global economy and has been growing continuously over the past decades. Africa which remains one of the poorest continents is benefiting from its appeal as a tourist destination, primarily for safaris and wildlife hunting, and has expanded its share of tourism arrivals and revenues. However, this positive momentum of tourism development is tainted by the debate on climate change which has put emissions and ecological impacts of economic activity in the centre of attention. In this light, the tourism industry has itself become criticised for the traces it leaves on the environment through increased traffic, the construction of hotel complexes, the squandering of water resources or the disposal of wastes. The sum of these factors defines the ecological footprint of tourism and highlights the often unsustainable use of earth's resources.

The Zero Footprint concept offers a great starting point for an investigation into the tourism industry in Africa and the effects on climate change. Starting with a theoretical introduction to the concept and its relevance for tourism, the discussion moves to the implications it could have on African tourism and the development of the continent. While highlighting the important balance which needs to be maintained between ecological concerns and chances for economic development, this chapter offers insights into means to reduce the footprint of tourism both locally at the micro-level through sustainable eco-friendly behaviour and globally at the macro-level through compensation for CO_2 emissions resulting from air travel.

2 Footprint Analysis and Its Application in Tourism

The Ecological Footprint (EF) is a concept which offers a succinct representation of the impact human consumption has on the global biosphere. Evidently, every form of human consumption requires in some way the availability of resources, the use of goods and the disposal of wastes, which originate from or affect nature.

The area of productive land and marine ecosystems available thus represent the basis for (human) livelihood and the EF is a means to analyse whether the use thereof for any given lifestyle is sustainable. As such the EF is an index which measures the resources used for production and those needed to assimilate the waste and emissions generated over a defined period of time, irrespective of the location of the land.[1] The use of resources is converted into the equivalent of area units which are utilized by an individual, a business, a nation or population.

The EF emerged as an indicator for sustainability in ecological economics during the early to mid-1990s. It was principally developed by William Rees and Mathis Wackernagel who framed the term with their publications and offered a first comprehensive EF analysis of human impact on the environment with their book "*Our Ecological Footprint: Reducing Human Impact on the Earth*" of 1996. Subsequently, the concept has gained popularity and various reports and articles are published regularly assessing the environmental impact of different countries, regions or economic sectors on global biocapacity.

The concept has also proven successful outside the academic realm for communicating the notion of ecological sustainability and the biological limits of the Earth, mostly though the work of non-governmental Organizations (NGOs). In spite of critics of the concept, the EF is nowadays recognized by many International Organizations such as the United Nations with its specialized agencies and environmental NGOs such as the WWF and Greenpeace as an effective accounting tool to assess the extent of humanity's use of biocapacity within in the context of sustainability.

One of the great benefits of the EF concept is its inclusive approach which offers the possibility of analyzing environmental impact across the entire value chain of economic activity. It creates a link between local use of natural resources and their global availability and also offers a means of assessing ecological impacts which are not restricted locally such as greenhouse gas (GHG) emissions.

While the EF indicator is primarily a tool for scientific analysis and calculations it is still a very accessible concept. Its iconicity is even reinforced by the notion of Zero Footprint (ZF) which is in a way a logical extension of the EF analysis, providing not merely an evaluation and estimation but offering a clear goal of completely reducing the unsustainable impact of human consumption of ecological resources and services. By itself, the ideal of zero footprint seems impossible given that every form of human activity results in a footprint on the environment which even expands with increasing living standards. Consequently, the ZF concept is intrinsically linked to offsetting mechanisms to redress negative impacts.

The notion of sustainability is a core value in many industries and due to its salience the footprint concept even gained its momentum in certain industrial sectors such as forestry or the mining business where corporations have voluntarily started to compensate for their business activities. The Rio Tinto Group, one of the world's

---

[1]   United Nations Environment Programme, *Global Environment Outlook – GEO4Environment for Development*, 2007.

leading mining companies, is a prime example of a business actor seeking to reduce its ecological footprint by incorporating sustainable development as a core
value. The economic activity of the group, though necessary to meet demands for
current livelihood, is detrimental to the environment. However, the company aims
at reducing their impact to a minimum and has even committed to making a "net
positive impact" on biodiversity at their operating sites. This progressive approach
reflects the zero footprint rationale and includes measures such as the
rehabilitation of land or offsetting the impact of deforestation on site with projects
seeking to protect and reforest natural spaces elsewhere.

Applying the ZF indicator to the tourism industry offers the possibility of
assessing the sustainability of tourism not only from a local perspective of
resources used in the country of destination, such as water consumption, waste
generation or the management of wildlife, but also includes the impact of travel
represented for example in kerosene emissions. Hence, in the context of the
current debate on climate change the ZF concept has gained even greater
importance as it allows for an assessment of the sustainability of an industry on a
global scale.

Tourism activities evidently put a strain on the biosphere in diverse ways.
There are the more direct effects such as the impacts of constructions on the
appearance and quality of the landscape, the damage resulting from emissions
generated from travel and energy generation and the contamination of water
resources. But tourism also puts pressure on wildlife through hunting and fishing,
which is a prime tourist attraction in the case of Africa. By means of noise
pollution or disposal of wastes large numbers of tourists can further present a
nuisance to the natural habitats they came to visit.

Awareness of the ecological footprints of tourism has meant that the industry
has come under criticism for their local as well as their global ecological damage.
The ZF concept offers a tool for understanding the diverse impact of tourist
activities in their totality, providing opportunities for minimizing the footprint by
optimizing the use of natural resources and accounting for the remaining impact
through compensation. Therefore, achieving a truly "zero" footprint in tourism
will only be possible by means of offsetting techniques which become ever more
important given the growth of the industry, especially in the light of GHG
emissions and climate change.

3   Tourism Development and Climate Change in Africa

3.1   Importance of Tourism for Africa

Over the last decades tourism has continuously increased worldwide and is now
one of the largest and fastest growing commercial sectors. The number of international tourists visiting Africa has increased as well. As a result of high annual
growth since the year 2000, the number of visitors to Africa rose to 44 million in

2007, according to the United Nations World Tourism Organization (UNWTO).[2] Although Africa currently accounts for just 6% of the global tourism market, this share has gradually increased from only 3.4% in 1990 and there is still potential for further growth. Moreover, income from tourism in Africa was about 25 billion Euros in 2007. However, these overall positive trends for tourism in Africa conceal the enormous differences throughout the continent: the four largest tourist magnets, South Africa, Morocco, Tunisia and Egypt, have a combined market share of 60% of tourism volume and income. For our purposes here we will now turn away from the Mediterranean-bordering countries and focus on tourism in Sub-Saharan Africa which differs drastically from that of North Africa.

Landscape, in terms of untouched, wild nature, is an African potential, whose value for tourism has only begun to be tapped. What makes African tourism destinations so special is possibly best displayed by how every child immediately associates elephants, lions and giraffes with Africa. The spectacular, large African mammals ("the big five") are what set Africa apart from other competing tourist destinations. This is why the classical photo-safari forms the backbone of most trips to Africa. The corresponding clichés ("Out of Africa", etc.) do not seem to detract from the allure of safari tourism. In fact, quite the opposite is true. They tend to be beneficial.

In addition to photo tourism, hunting-tourism still has its place in Africa. Passionate big game hunters from western countries, as well as those from Asia and the Gulf States, are willing to spend fortunes for the chance to shoot a lion; naturally, an antelope can be had for less. A developed market for hunting-tourism in southern and eastern Africa allows every client hunter to shoot the trophies they are prepared to pay for. The meat of the hunted animal is generally turned over to local population and temporary jobs for bearers, trackers, cooks and others are created. The challenge, from a development point of view, is to ensure that the largest possible amount of the income generated by hunting-tourism is left to local communities.

Programmes in which local communities are responsible for the preservation and care of wildlife (Community Based Natural Resource Management, CBNRM) have proven especially promising. In return the communities receive the right to sell hunting licences and thereby profit from the maintenance of wildlife populations. This raises the communities' appreciation for wild animals and demonstrably contributes to increases in wildlife populations. Because of this, competent partners, e.g., well-known NGOs, such as WWF, Conservation International and others, have helped to establish these types of programmes together with the relevant government services, in countries like Namibia, Botswana and Tanzania (formerly in Zimbabwe as well).

The wild animal safaris that make African tourism stand out can almost always be combined with a trip to the ocean. The islands and beaches along the eastern African coast can stand up to comparison with any of the better-known beach

---

[2] United Nations World Tourism Organization, *Tourism Highlights*, 2008.

tourism destinations (Maldives, Thailand, etc.). The range of beach tourism stretches from "hotel castles" along the Red Sea, which have already found their place in the catalogues of European travel agencies, to close-to-nature camps in Mozambique, which are primarily visited by individual tourists, and to exclusive resorts with all imaginable luxuries, for example, in the Seychelles. Hence, beach tourism in all its various forms can **also** be found in Africa. However, the most extraordinary and unique tourism product of Africa is and remains the wild animal safari.

The important employment and income effects that can be attained through the marketing of the African landscape and nature are best seen in southern Africa. Given the fact that tourism is one of the strongest growing commercial sectors worldwide, even countries that currently lay off the beaten path will have enormous development chances in the future. Africa is heavily reliant on the money brought in by tourists.

3.2   Effects of Climate Change in Africa

We all know that Africa emits only a small amount of CO_2. Reliable studies show that Africa accounts for only 3-5% of total worldwide emissions. Therefore, Africa is affected by climate change, but has barely contributed to its emergence. With the exception of greenhouse gases released from fire clearing of forests and tree savannahs in order to make room for new crop fields and the emissions of several industrialising countries, such as Egypt, South Africa and the Maghreb states, the emissions of the continent have been negligible. This statement could be made even more drastic if one only observed Sub-Saharan Africa and per capita emissions.

It is also uncontested that Africa will suffer more from the effects of climate change than any other continent. When precipitation grows more irregular and incalculable as a result of climate change, extreme events will occur more often: droughts and floods will become more common occurrences on the continent and will have catastrophic consequences for the people who, due to expanding poverty and an extreme reliance on agriculture, are substantially more susceptible to climate fluctuations than those living in industrial nations. This is one reason why it is necessary to adapt agriculture, which in African countries remains a basis of life for most of the people there. The necessary measures are mostly known and have largely been tested. Agricultural research (e.g., the development of drought resistant crops), efficient use of the key resource water (e.g., through irrigation systems) or better links between small farmers and their markets (e.g., through the construction of rural roads) are only some examples.

Longer lasting dry spells lead to the marginalisation or even abandonment of previously productive agricultural areas. The affected farmers naturally try to recover from these losses in other locations and seek out new fields, even in protected areas. Although this is illegal, it can hardly be stopped as the rural population is often driven by shear need. Because of this chain of events, natural African treasures are in great danger during times of climate change. At the same time, these same natural treasures are the tourist magnets of Africa.

3.3   Consequences for the Zero Footprint Concept

It is evident that in order to maintain long-term tourist potential, nature will need to be handled carefully and ecologically sensitively. Sustainability is critical. The concept comes from the forestry industry and states: don't cut down more trees than can grow back and plant new trees so that coming generations will also have access to wood. This principle can be readily carried over to the tourist potential of Africa. Whether a game reserve can be sustained for a long period is easily measured, for example by changes in the density and diversity of wildlife over time. This kind of sustainable nature management will ensure that tourism does not leave behind a negative footprint, at least at the local level, for example, in a particular national park. Sustainable nature management within the framework of development cooperation has already been encouraged long before the recent climate debate. For many countries in Africa it would be a major success, if this goal could be reached.

The question about the **global** footprint of photo, hunting or beach tourism remains unanswered. In particular, this will require that the emission of climate damaging gases, which are emitted in considerable amounts during any long-distance flight, have to be considered as well. To be exact a complete climate balance for every tourist would need to be generated in which the total environmental wastage of their holiday trip is taken into account and includes, for example, everything from their transfer from the airport to their hotel to the water needed to wash their towels. In order to meet the requirements of zero footprint tourism, the tourism branch would need to make expensive investments in emission reductions on site or provide considerable compensation. It is clear that such a "total climate score" would be a heavy burden for African tourist destinations.

This is where the central question arises. Can countries that are very poor and have not contributed to climate change be expected to put climate protection above the need to improve the standards of living for the vast majority of their people?

Both climate protection and poverty reduction are desirable goals, but they are competing with one another. One must assess in how far a continuation of poverty is acceptable in order to protect the climate or, on the flip side, how much global warming is tolerable to allow the development of poorer countries. This is also a moral question.

When it can be assured that an appropriate amount of tourism income is left to the poorer local communities, then a tourism project on the coast or in a national park should not fail due to climate protection. African countries must profit from tourism as a fast growing commercial branch. Very high climate obligations right from the start should not rob them of one of the few viable development paths. Others should be doing more for climate protection than Africans. Pressure from customers and non-governmental organisations will need to increase so that airlines, tour operators and hotel complexes take a vested interest in financing climate protection activities, in order to compensate for the GHG emissions originating from tourism.

4   The Potential of Compensation Payments

Zero footprint tourism still remains an ideal in Africa, or in any other region of the world for that matter, but there are examples of initiatives which are moving in the direction of making tourism more sustainable at the micro and the macro-level of impact. The ZF concept exemplifies that even with the best optimization techniques for reducing the damage to the environment locally, truly sustainable tourism can only be achieved by coupling ecological improvements with compensation measures.

It is especially in the context of the climate change debate that compensation has increasingly gained popularity, often focusing offsetting activities on balancing GHG emissions. Offsetting offers a chance for emitting companies, such as airline carriers, hotel chains and tour operators, to limit their footprint either by directly financing compensation projects in afforestation for example or by means of buying emission certificates either on the regulated or on the voluntary market.

Tourism companies seeking to compensate their emissions and reduce their footprint can draw upon the expertise of procurement programmes such as the KfW Carbon Fund which buys emission certificates according to the flexible project-based mechanisms of the Kyoto Protocol (Clean Development Mechanism, Joint Implementation). The KfW Carbon Fund operates on the regulated markets. The eligibility within the EU Emissions Trading System (EU-ETS) of the acquired certificates is an important prerequisite for the selection of projects. This guarantees a high standard of the activities supported, but compensation via the voluntary market has the benefit of offering more flexibility and diversity in projects allowing for trial and error.

Activities seeking to avoid deforestation, or regenerate endangered and degraded forests and land also reduce GHGs. While most of these projects are not recognised under the EU-ETS or Kyoto system, they produce Verified Emission Reductions (VERs). These carbon credits can also be sold to tourism companies or individual tourists wishing to voluntarily offset their emissions. When the possibilities of directly minimizing the footprint of tourism such as improvements in energy efficiency or resources management are exhausted, purchasing VERs is an effective means of offsetting the remaining impact. Projects in the voluntary market are usually on a smaller-scale with great potential for offsetting emission, but lacking the capacity to withstand the costs of compliance with Kyoto or EU-ETS demands (certifiers, validators, consultants etc.).

The voluntary market is thus a place for testing and trying new innovative approaches in the field of reducing emissions. However, the downside of this trend-setting potential of the voluntary market is the danger of greenwashing, as the market lacks standards for assessing the effectiveness of projects. This implies that activities which are promoted as carbon reducing initiatives run the risk of being mismanaged or potentially even intentionally misused which calls for a necessary degree of caution when assessing the projects intended to reduce footprints to make sure the rotten apples do not spoil the barrel.

The variations of compensation payments may seem bewildering, but they offer the chance of coupling ecological considerations with development concerns, because these voluntary carbon credits can be generated through projects which are supported by development cooperation institutions and actors, such as KfW Development Bank. German financial cooperation for instance is funding a project in Madagascar aiming at the reforestation of more than 9,000 ha through sustainable forest management. Similar projects exist in China and Vietnam, which again exemplifies the beauty of the ZF concept, because compensation is not locally restricted but can occur globally. The VERs gained through carbon sequestration by these projects offer the afforesting cultivators the chance of earning higher revenues from forestry creating further incentives for the reduction of GHGs and providing development opportunities. These projects serve the environment and help reduce footprints, while also benefiting the reduction of poverty.

Looking at the application of offsetting mechanisms in the tourism sector, exemplifies that there is still a need for tourism companies and customers taking the initiative to achieve zero footprints. An example of a tourism actor engaging directly in footprint reducing activities is the online holiday agency TravelRepublic. On the one hand, they are giving their clients the opportunity to voluntarily neutralize the carbon emissions generated from their flights by donating money to the World Land Trust charity for the regeneration of endangered rainforests, helping to offset emissions as well as protect biodiversity. On the other hand, the company itself is greening its image and pledges to save a tree in the Amazon for every return flight booked with them.

Through increased public pressure by NGOs and consumers, such climate-friendly initiatives will become more important for the tourism sector to preserve a positive public image and maintain the attractiveness of the destinations. However, achieving zero footprint in tourism will only become viable if tourists themselves assume part of the financial burden which results from offsetting. At least in the case of tourists in Africa, the ability to pay of tourists from industrialized countries should be tapped as long as it does not impede on tourism development and revenues. The benefit of compensation payments for the goal of zero footprint tourism is evident, because it factors in the global effects of environmental damage, but it does not relieve the destination of their responsibility to sustainably manage the environment.

5   Case Studies: Mozambique and Madagascar

There are currently many publications that cover the burden on the environment caused by long-distance travelling. These articles also include possible measures for balancing the effects. But what about the tourist infrastructure on location, what can be done there to minimise the impact on the environment and simultaneously help economic growth and increase standards of living for the people of the host country? What can partner countries, tourist operators, lodging providers and the tourists do to minimise the ecological footprint of tourism?

One example of a national approach comes from Madagascar.[3] Here the government has set forth principles for ecologically oriented tourism in their 5-year Madagascar Action Plan (MAP):

- Ensure environmental sustainability by adopting strategies for sustainable development and the protection of natural resources

- Use resources intelligently and productively to minimize the loss and maximize the gain

In order for these political formulations to take shape, MAP includes the following operation guidelines for implementation:

- Identify and launch new tourist sites and products

- Promote the Destination Madagascar as a superior and unique ecotourism destination

- Establish a national ecotourism framework and strategy to contribute to the protection and promotion of the environment and to ensure "eco-eco" harmonisation (economic – ecological).

- Establish an ecotourism policy, charter, code that states the vision, the commitment, the values and the approach for the promotion and implementation of ecotourism throughout the country

- Establish special zones for ecotourism

Madagascar is definitely playing the ecological card strongly, in order to establish itself in the international long-distance travel market. Ecological and nature-aware behaviour on the part of the travel destination has become a targeted marketing point. This has increased awareness for the country among a growing number of international tourists and the growth rate for visitor volume has stayed above 10% for years. However, this also means that tourists visiting the country also expect a

---

[3]   Pawliczek, M. and Mehta, H., "Ecotourism in Madagascar: How a Sleeping Beauty is Finally Awakening", in Spenceley, A. (ed.), *Responsible Tourism: Critical Issues for Conservation and Development*, 2008.

certain level of ecological orientation in the country's tourism infrastructure. They are particularly sensitive about so-called "greenwashing". For this reason, lodgings that label themselves as an "ecolodge", for example, without knowing and/or fulfilling the appropriate criteria, are to be viewed critically and are damaging to the reputation of the tourist destination.

The following criteria, defined by the International Ecotourism Society in 2002, should be applicable in Madagascar in the future. In order to assume the label "ecolodge", at least 5 of the following points must be fulfilled:

- Help in the conservation of the surrounding flora and fauna

- Have minimal impact on the natural surroundings during construction

- Fit into its specific physical and cultural contexts through careful attention to form, landscaping and colour, as well as the use of vernacular architecture

- Use alternative, sustainable means of water acquisition and reduce water consumption

- Provide for the careful handling and disposal of solid waste and sewage

- Meet its energy needs through passive design and renewable energy sources

- Use traditional building technology and materials wherever possible and combine these with their modern counterparts for greater sustainability

- Endeavour to work together with local community

- Offer interpretive programmes to educate both tourists and local communities together with employees

- Contribute to sustainable local development through education programmes and research[4]

In Madagascar, according to Pawliczek, lodges that label themselves "ecolodges", either under false pretences or simply because they are unaware of the requirements, are currently popping up like weeds. But an ever-growing number of businesses, such as the tour operator "Boogie Pilgrim" together with the NGO "Fanamby", the "Princes Bora Lodge", the "Tsara Guest House" etc., are already implementing the mentioned criteria.

One reason why these ambitious criteria are not being adequately met by other operators, particularly from the structural point of view, is because of the additional costs involved that the private lodge operators in Madagascar are unable to cover alone. The island's banking sector is averse to risks and sees the tourism sector as particularly hazardous, because of its high volatility and susceptibility to external shocks. For this reason German financial cooperation has created a capital

---

[4]   Mehta, H. 2002 in Pawliczek, M. and Mehta, H., 2008.

fund through the KfW Development Bank together with the parastatal Madagascan national park agency Business Partners International and the IFC. This fund shall enable tourism businesses to manage the additional costs, and therewith allow them to build tourism camps and simple lodges in line with ecological standards on the edges of national parks. Targeted credit lines are thus an important tool in concertedly minimising the ecological footprint of tourism in Africa.

Another prime example of how the mentioned criteria can be implemented is the Guludo-Resort in Mozambique, whose owner, Amy Carter-James, displayed the approach at the ITB (International Tourism Exchange) African Day in 2008. The Guludo-Resort resulted from the threefold desire to make money, reduce poverty and protect the environment, which is why the business model includes a hotel operation (lodge) on one side and a foundation (Nema Foundation) on the other.

The products offered in the lodge are primarily Fair-Trade; an advantageous decision for the environment, the profitability of the business and the producers of the raw materials. The poverty-reducing approach of the Guludo-Resort is anchored in the fact that investments in tourism lodges in extremely poor regions are being realised. Simply setting up a lodge of this type provides substantial economic benefits for the people living there, through the creation of jobs and marketing possibilities for local products. Established community groups are integrated from the earliest construction phases and can win additional income, for example, when groups of women come together to produce ceramic floor tiles.

The design and construction of the lodges together with the configuration of the tourism operation are the most important aspects, as far as environment protection is concerned. Traditional customs and habits of the local culture are respected and mostly local materials are used in the construction of buildings. Additionally, the lodges are managed so as to have a minimum impact on the environment, by teaching better fishing techniques to the local community and because tourist activities in the ocean have been conceived and realised by marine biologists.

Income from the lodge covers the operational costs of the Nema Foundation, ensuring that 100% of donations made by tourists go directly towards development projects at location. The foundation has been active for about three years now and their projects have had a tremendous effect in fighting poverty and protecting the environment. With help from tourism in the Guludo-Resort in Mozambique 15,000 people have been provided with potable water, mosquito nets have been distributed, malaria training conducted with 12,000 people, two primary schools and a secondary school have been built, small local businesses have been founded and ocean and forest protection projects have been financed.

The Nema Foundation's most current project financed with money from tourists is a CO_2 emission-reduction programme. Despite its best efforts, the Guludo-Resort still displays the difficulty involved in creating a tourism resort without an ecological footprint, especially considering that most of their guests arrive at the lodge after long-distance flights. As a solution, the foundation offers guests the chance to compensate their share of CO_2 emissions by making donations to the Nema Foundation. The donations are turned over to a local coalition who, for example, uses it to replant and protect forests. Another portion of the donations go towards small health, education and environment protection projects on location. The achievements of these projects are the result of the success of tourism and combine economic and ecological goals.

6   Conclusion

Ultimately, tourism activities are not only important for the economic welfare of Africa, but they can have a positive impact on the development of wildlife numbers and natural diversity on the African continent, if the revenues are fairly distributed. It is in the self-interest of African countries to remain an attractive tourism destination and sustainably care for their richness in tourism assets. The criticised ecological footprint of tourism can thus be minimized through adapted policies, attitudes and behaviour, as the examples from Madagascar and Mozambique show. Ecologically-friendly lodges have been created, targeting a more high-end tourism able to afford the additional costs of sustainability and in return offering the experience of a more traditional lifestyle and culture in the country of destination. The detrimental climate impact of long-haul flights can be reduced by targeted compensation payments or the funding of afforestation projects for instance. However, to achieve the ideal of zero footprint tourism, the markets for emissions trading must be further developed in the post-Kyoto era in a way which ensures that Africa's development is not impeded from catching up with the industrialized countries. Applying the ZF concept to tourism in Africa offers the possibility of joining ecological concerns of sustainability with economical endeavours. However, the often conflicting ambitions of climate change impacts and tourism potential need to be thus balanced, not to hinder African development.

Aviation Management

Kerosene's Price Impact on Air Travel Demand: A Cause-and-Effect Chain

Richard Klophaus

Abstract

This paper examines the impact of rising fuel prices on future air traffic. Using route and carrier specific data the short-term impact of higher fuel prices on airline operating costs, passenger fares and demand for short-haul and long-haul services is analyzed. Results suggest that the air traffic growth, constrained by scarcity of kerosene, will be much lower – or even negative – than unconstrained air traffic growth. Services offered by low-cost carriers and long-haul services are most adversely affected. Further, a strong increase in fuel prices outweighs the impact of proposed emission trading systems for the aviation industry.

1 Introduction

The aviation industry is probably the economic sector which is most depending on fossil fuels besides the petroleum and petrochemical industry itself. Today, commercial aviation is characterized by growing passenger numbers and cargo volumes as well as expanding airport and airline capacities. Long-term forecasts of continuously growing air traffic are issued by manufacturers of commercial jetliners like Airbus and Boeing and proliferated by other sources including public agencies and academia. Based on these forecasts the demand for kerosene is bound to grow.

A rising oil demand is forecast aside from the transportation sector. The reference case of the Annual Energy Outlook 2008 released by the Energy Information Administration of the US Department of Energy projects a growth of oil demand from 83.6 million barrels per day (mb/d) in 2005 to 112.5 mb/d in 2030 (EIA 2008). Annual average oil prices increased in every year since 2003 due to growing worldwide demand. By the time peak oil is reached and half of the global oil resources are exploited, costs for oil extraction will rise and keeping up the production level will become increasingly difficult. The depletion of the world's oil reserves results in an upward price trend for crude oil and also for its refinery products such as kerosene.

This paper investigates the economic impact of higher oil prices on fuel costs, air fares and air passenger demand. The analysis uses the methodology developed for quantifying the impact of emission trading on aircraft operators. With regard to the relationship between kerosene prices and airlines' fuel costs different fuel hedging scenarios are considered. The paper indicates that the rate of air traffic growth constrained by scarcity of kerosene is much lower – and may even be negative – than unconstrained air traffic growth, especially leading to a strong reduction of demand for leisure traffic and long-haul services.

The paper is structured as follows: Section 2 contains considerations on peak oil including an overview of predictions on the time of global oil production peak. Section 3 examines, assuming an horizon of one year, the short-term economic impact of higher fuel prices on airline costs, ticket prices and passenger demand for short-haul and long-haul services. Short-haul is further differentiated into routes operated by full service network carriers (FSNCs) and low-cost carriers (LCCs). In the long run the aviation industry has to look beyond the fuel-efficient '3 liter aircraft' and search for new groundbreaking ways to become less dependent on fossil fuels. Hence, Section 4 gives an overview of current research directions in the fields of future aircraft technology and evaluates potential alternative fuels to kerosene. The closing Section 5 summarizes the paper's results and concludes that peak oil has the potential to stop and even reverse long-term air traffic growth.

2   Peak Oil and Future Fuel Prices

At the end of Feb. 2008 the spot price of Brent-Europe crude oil reached for the first time $100/b (Fig. 1). This reflects that world oil demand has continued to grow faster than oil supply but also ongoing geopolitical risks, OECD inventory tightness, worldwide refining bottlenecks and speculative trading. Prices rose further in 2008, passing $140/b in July, well above the historical inflation-adjusted record price set in the early 1980s following the Iranian Revolution and the beginning of the Iran-Iraq war.

Kerosene is produced by distilling crude oil. Hence, its price is closely linked to crude oil prices. At the end of Sep. 2008 the spot price for kerosene-type jet fuel in Rotterdam was about 305 Cents per gallon. This translates into close to $128/b (Fig. 1) in comparison to $97b for Brent-Europe. The spread of approx. $30/b on the spot price of Brent-Europe crude oil reflects the gross refining margin. The annual average refining margin steadily increased from its value of just $4/b in 2002.

Fuel outranked labor as largest single cost item in the global airline industry and accounted for 29% of total operating costs in 2007 compared to 13% in 2001 (IATA 2008). The rise in fuel costs reflects a sharp increase in the price of crude

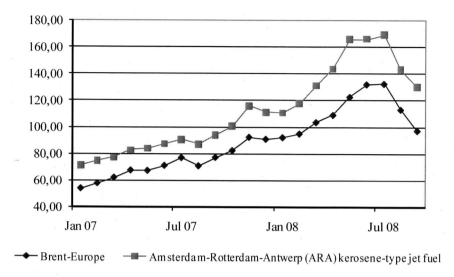

Fig. 1. Spot price (FOB) Brent-Europe and kerosene-type jet fuel. Source: EIA 2008

oil but also a widening in the refining margin. Counteracting soaring fuel costs airlines intensified their efforts to improve fuel efficiency and to obtain cost savings in non-fuel cost items. In particular, labor productivity has improved resulting in a falling labor share of airline operating costs. The 29% fuel share of total operating costs calculated by IATA rests upon a price of crude oil of $73/b. With spot prices of crude oil roaring past $100/b and beyond the share of fuel costs further increases in the airline industry even if fuel hedging contracts lock a percentage of the fuel purchases at lower prices. Airlines react by increasing average prices for passenger tickets and rates for air cargo, e.g. by raising fuel surcharges on fares.

Crude oil prices react to the balance of demand and supply. Hence, the spiking of fuel prices creates concerns about a global shortage of future oil supplies. If actors in the oil market expect a shortage of oil supplies, oil prices increase before a shortage actually occurs. This is reflected in contracts for future deliveries of crude oil, called futures. For example, in Oct. 2007 prices of crude oil futures soared to all-time highs after the Energy Information Administration (EIA) indicated a drop in commercial US crude inventories to the lowest level in two years. EIA providing the official energy statistics from the US government publishes an International Energy Outlook (EIA 2008). In the so-called reference case of its most recent outlook, EIA projects a growth of world consumption of petroleum products from 84 mb/d in 2005 to 113 mb/d in 2030, an average annual growth rate of 1.1%. A disproportionately high increase in consumption is projected for non-OECD countries in Asia and the Middle East, where strong economic growth and rapidly expanding transportation use is expected.

In addition to the reference case with a world oil price of $113/b in 2030, EIA (2008) includes a high oil price case with $186/b leading to a consumption of

99mb/d in 2030. This indicates that long-term demand is relatively inelastic to oil price changes. It is a question whether the suggested lack of demand elasticity remains a valid proposition once production of crude oil falls short of demand due to finite oil reserves. If global crude oil production cannot be increased even with mounting oil prices there has to be a pronounced demand adjustment.

The US Government Accountability Office (GAO 2007) examined more than twenty studies on the timing of the peak in oil production conducted by government authorities, oil companies and oil experts (Fig. 2). According to this meta-analysis most studies estimate peak oil sometime between now and 2040. The range of estimates on the timing of peak oil is wide due to multiple and uncertain factors including (1) the amount of oil still in the ground, (2) technological, cost and environmental challenges to produce that oil, (3) political and investment conditions in countries where oil is located and (4) the future global demand for oil. Some of the studies cited by GAO consider only the peak in conventional oil, while other studies include non-conventional sources of oil – oil sands, heavy and extra-heavy oil deposits and oil shale. The production process of oil from non-conventional sources is more costly, uses larger amounts of energy and presents environmental challenges.

The oil production from conventional sources in some Non-OPEC regions like North America and North Sea has already peaked or will do in the near future. Approximately 70% of the estimated remaining global oil reserves are located in politically insecure regions respectively are kept under OPEC control (EIA 2008). OPEC statements concerning strategic oil reserves may be questioned. Oil production represents a major sector of economy in OPEC countries, and the admission of declining oil reserves harms their financial standing and political importance. The number of discovered oil fields decreases year by year. About 42,000 oil fields have been discovered until today, the 400 largest represent about 75% of global oil reserves. The annual worldwide crude oil consumption exceeds the amount of discovered reserves since 1981. The predominant part of extracted crude oil nowadays derives from oil fields discovered in the 1970s (Deutsche Lufthansa 2007).

Finite oil resources and global economic growth lead to an upward trend for crude oil prices. However, due to multiple and uncertain factors concerning near-term and long-term oil production and the future development of global oil demand it is not surprising that forecasts on future prices show a wide range. The following section of this paper analyzes the impact of a 50% differential in spot prices for crude oil for a one-year planning period, more specifically between $83/b as the one-year forecast released by Air France-KLM (AF-KLM) in Oct. 2007 and $125/b. The second figure is neither the actual spot price in Oct. 2008 nor another oil price forecast for the following months but only serves as starting point for the analysis of the possible short-term impact of higher kerosene prices on commercial aviation. $125/b is beyond the oil price ranges of most recently published short-term forecasts for 2009. However, most short-term oil price forecasts published in recent years underestimated the actual oil price development.

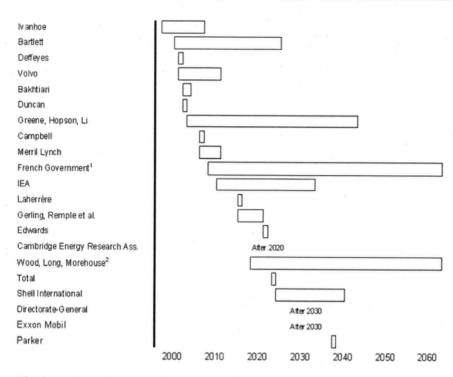

1) End of estimated timespan is out of scale (2125)
2) End of estimated timespan is out of scale (2115)

Fig. 2. Estimates of the timing of peak oil (GAO 2007)

3   Economic Impact of Soaring Kerosene Prices

This section considers the short-term response of passenger demand for air travel resulting from the cause-and-effect chain depicted in Fig. 3. This cause-and-effect chain is referred to as direct impact of higher fuel prices. In addition, there are potential indirect impacts of soaring fuel prices, for example, a reduction in air travel demand resulting from lower disposable income of households or an increase in airlines' operating costs for other cost items than fuel due to inflation.

The paper assumes no disruptions in normal economic activity and that the overall political and economic setting for commercial aviation remains intact. The time horizon is only one year allowing to differentiate a scenario with a high level of airline fuel hedging and a scenario with no fuel hedging a year ahead. This short-term approach justifies not to account for fuel efficiency measures and also to use current operational and cost data.

The volatility of kerosene prices is an important issue for the airline industry. In 2006, the fuel consumption of Lufthansa (LH) amounted to 6,940,587 tons equiva-

Fig. 3. Direct economic impact of higher fuel prices

lent to 54,564,363 barrels (1 barrel = 159 liter, 1 liter kerosene = 0.8 kg) (Deutsche Lufthansa 2007). Without fuel hedging a fuel price rise of $1 a barrel increases LH's operating costs by more than $50 millions. Fuel hedging is often touted as the solution to this problem.

Fuel hedging means stabilizing fuel costs by locking in the costs of future fuel purchases to protect against sudden cost increases from rising fuel prices. However, fuel hedging also prevents savings from decreasing fuel prices and might even lack a theoretical justification (Morrel & Swan 2006). In practice, fuel hedging strategies vary significantly between airlines, some opting to hedge their entire fuel needs, while others leave themselves exposed to fluctuations in fuel costs. The lack of fuel hedging might not be strategy-driven but simply the result of insufficient cash or credit. According to Morrell and Swan, most airlines typically hedge between one- and two-thirds of fuel costs and look forward six months in their hedging, with few hedges more than a year ahead. Hence, the fuel hedging policy of AF-KLM with regard to the time period covered seems to be rather exceptional (Table 1).

In Oct. 2007 AF-KLM forecasted a spot price of $83/b in 2008-09. The average fuel price for AF-KLM is locked with fuel-hedging contracts that secure 67% of the airline's fuel requirements in 2008-09. Even for 2010-11 31% of the fuel needs are hedged. The average hedged price in 2007-08 with $62/b was only two thirds the spot price of crude oil at end of Oct. 2007 of more than $90/b. By using the

Table 1. AF-KLM fuel hedging policy. Source: www.airfranceklm-finance.com

Year	2007-08	2008-09	2009-10	2010-11
Forecasted spot price (Brent, $/b)	79	83	79	78
Hedged consumption (%)	77	67	51	31
Average hedged price (Brent, $/b)	62	61	68	69
Final average price (Brent, $/b)	66	68	73	75

futures markets AF-KLM managed to soften the effect of higher fuel prices but still increased its fuel surcharges on air fares. From the information publicly available, AF-KLM hedges only the crude oil price. This leaves the price difference between crude oil and kerosene uncovered. The kerosene price is driven by crude oil price developments but is also influenced by other specifics, especially refinery capacities and price switches between diverse oil products.

This paper analyzes the impact of a 50% differential in spot prices for crude oil for a planning period of one year, more specifically between $83/b as the one-year forecast released by AF-KLM in Oct. 2007 and $125/b. This represents the same rate increase in the year to Oct. 2007. As already stated in Section 2 the $125/b is not another oil price forecast but only serves as starting point for the analysis of the short-term impact of higher kerosene prices on commercial aviation. Additionally, a constant refining margin of $30/b is assumed. Hence, all further calculations are based on spot prices for kerosene-type jet fuel of $113/b and $155/b. This fuel price of $113/b represents the scenario forecasted by AF-KLM in Oct. 2007 for 2008-09.

All airlines are confronted with volatility of fuel prices. Besides the structure of fleet and network different fuel hedging policies lead to a varying effect on fuel costs among airlines. In Europe, network carriers like AF-KLM and LH pass along higher fuel costs to passengers through higher ticket prices with changing fuel surcharges added to air fares. Ryanair (FR) and other European LCCs do not add fuel surcharges but increase the average fare level. The following calculations estimate the increase in ticket prices and the resulting changes in passenger demand due to rising fuel prices. As reference value to measure fuel price increases the final average price of $68/b forecasted by AF-KLM for 2008-09 by end of Oct. 2007 is used (Table 1). This fuel price is further on referred to as AF-KLM base. As no similar information about spot price forecasts, hedged and average fuel prices for LH and FR has been available, this paper assumes that the AF-KLM base is also valid for LH and FR.

The airline's fuel costs ($/liter kerosene) at a given future time based on the future spot crude oil price, the airline's hedged consumption, the average hedged price and the refiner margin can be calculated as follows:

$$C = ([\alpha \cdot p_h + (1-\alpha) \cdot p_s] + c_r)/159 \tag{1}$$

with

C Future fuel costs ($/liter kerosene),
α Share of fuel consumption hedged (%),
p_h Average hedged crude oil price ($/b),
p_s Future spot crude oil price ($/b),
c_r Gross refining margin ($/b).

Assuming the AF-KLM forecast of a spot price p_h= $83/b for crude oil and the airline's fuel hedging policy (α=0.67; p_h=$61) for 2008-09 (see Table 1) and a refiner margin c_r=\$30/b, the fuels costs (\$/liter kerosene) for AF-KLM resulting from (1) amount to C= 0.62. In comparison, a spot price p_h= \$125/b for crude oil in 2008-09, i.e. 50% higher than forecasted by AF-KLM in Oct. 2007, leads to C=0.71 other things being equal. Hence, with a high level of fuel hedging fuel costs are approx. 15% higher than forecasted. In a scenario with fuel hedging contracts running out or no airline fuel hedging a year ahead (α =0) future fuel costs even rise to C=0.97. To account for soaring fuel costs, airlines like AF-KLM increase fuel surcharges on passenger tickets.

The economic impact of higher fuel costs passed on to passengers via higher ticket prices is investigated for exemplary routes (Table 2). Based on the operational data provided by Scheelhaase, Grimme & Schaefer (2007) following routes are analyzed:

- Frankfurt (FRA) – London-Heathrow (LHR) served by LH.

- Hahn (HHN) – London-Stansted (STN) served by FR.

- LH-operated intercontinental route FRA – Singapore (SIN).

For each route the analysis differentiates the impact of fuel price increases with fuel hedging (α=0.67) and without fuel hedging (α=0). The results are compared with the AF-KLM base of a final average crude oil price of \$68/b. Due to lacking airline-specific information it is assumed that any increase in fuel costs in excess of the AF-KLM base is fully passed on to passengers via higher ticket prices, that is fuel cost increase equals ticket price increase. FR has a no fuel surcharge policy and accommodates higher fuel prices by increasing average ticket prices. LH increases its fuel surcharge on air fares. In July 2008 LH's fuel surcharge on long-haul tickets amounted to €97 per sector and for short-haul tickets €27 per sector. Since 2007 LH raised its fuel surcharge several times in lockstep with soaring oil prices until the record level reached in July 2008.

Table 2 shows the fuel consumption and the average passenger number per flight for the three selected exemplary routes. Based on this data, the route-specific future fuel costs per passenger can be calculated for AF-KLM base as reference value (C=0.62) and two fuel hedging scenarios (C=0.71 and C=0.97) as follows:

$$c_{pax} = C \cdot \frac{k}{n} \tag{2}$$

with

c_{pax} Future fuel costs per passenger (\$/PAX),
C Future fuel costs (\$/liter kerosene),
k Fuel consumption per flight (liter kerosene),
n Average passenger number per flight (PAX).

As FR and LH denominate their ticket prices in Euro, fuel costs need to be converted into Euro as well. Table 2 uses the Dollar/Euro exchange rate €1 = \$1.45 valid end of Sep. 2008.

Table 2. Increases in fuel costs and ticket prices due to higher kerosene prices

Route (carrier)		HHN-STN (FR)	FRA-LHR (LH)	FRA-SIN (LH)
	Distance flown	572	695	10,603
	Aircraft type	B 737-800	A 321-100	A 340-300
	No. of seats	189	182	247
Operational data	Avg. seat load factor	76.1%	66.9%	80.5%
	Avg. no. of passengers	144	122	199
	Fuel consumption (liter kerosene)	3,250	4,125	107,500
	AF-KLM base	14.0	21.0	334.9
Fuel costs (c_{pax})	$\alpha = 0.67$	16.0	24.0	383.5
	$\alpha = 0$	21.9	32.8	524.0
	AF-KLM base	9.7	14.5	231.0
Fuel costs (€/PAX)	$\alpha = 0.67$	11.0	16.6	264.5
	$\alpha = 0$	15.1	22.6	361.4
Avg. ticket price (per sector, €/PAX)		44	136	602
Abs. price increase (€/PAX)	$\alpha = 0.67$	1.3	2.1	33.5
	$\alpha = 0$	5.4	8.1	130.4
Rel. price increase	$\alpha = 0.67$	3.0%	1.5%	5.6%
	$\alpha = 0$	12.3%	6.0%	21.7%

The short-term impact of soaring kerosene prices on fuel costs and ticket prices depicted in Table 2 remains relatively moderate as long as fuel hedging by airlines mitigates fuel price increases. In absolute terms the price increase for the two short-haul routes is €1.3 (HHN-STN) and €2.1 (FRA-LHR) corresponding to a relative price increase per passenger and sector of 3.0% and 1.5% respectively. The higher price increase in per cent for FR results from its significantly lower average ticket price per sector and passenger compared to LH which cannot be compensated by FR's shorter flight distance and the higher average number of passengers per flight. For the long-haul route FRA-SIN the impact is already more pronounced, with an absolute price increase of €33.5 corresponding to 5.6% in relative terms. It should be noted that even in a scenario with fuel hedging (α=0.67) the impact of increasing fuel prices is higher than the financial burden due to the introduction of the emission trading scheme as proposed by the European Commission (Scheelhaase & Grimme 2007).

For the scenario with fuel hedging contracts running out or no airline fuel hedging (α=0) the impact is much stronger. In relative terms, LH's short-haul operation is less affected than the operation of FR. The average ticket price sold by LH on FRA-LHR rises by 6.0% (€8.1) and FR's price on HHN-STN by 12.3% (€5.4). The impact on surcharges on long-haul traffic largely exceeds the impact on short-haul traffic. For FRA-SIN operated by LH an additional fuel surcharge of €130.4

would occur. Based on an average fare per passenger and sector of €602, the additional fuel surcharge represents a relative fare increase of 21.7%.

€130.4 is the additional fuel surcharge calculated for LH resulting from assuming spot prices for crude oil to rise by 50% to $125/b compared to $83/b and no softening of spot prices by fuel hedging. In principle, this paper equates short-term with a one-year horizon. In its hedging practice, LH hedges up to 90% of its planned fuel requirement on a revolving basis over a period of 24 months. Even 70% is hedged one year ahead (www.lufthansa-financials.de). LH reduces its hedging ratio from this share each month by 5% leading to a growing exposure to fluctuations in fuel prices after one year. Hence, our results seem to overestimate the short-term impact of rising fuel costs on LH's fuel surcharge. However, FR has complained to the EU about abusive increases in fuel surcharges based on spot prices on the global crude oil markets rather than hedged prices. According to FR, carriers like LH do not only increase ticket prices for passengers in lockstep with their higher fuel costs but even beyond their additional fuel costs.

As a result of shifting costs to passengers via higher ticket prices, demand for flights decreases. Table 3 shows how passenger demand reacts to higher ticket prices. The average price elasticity for short-haul leisure and business demand as well as for long-haul leisure and business demand is taken from a synoptic study (Gillen, Morrison & Stewart 2004), the shares of business travelers are adopted from Scheelhaase, Grimme & Schaefer (2007) and the relative increases in ticket prices from Table 2.

In the fuel hedging scenario (α=0.67), the estimated change in passenger demand for HHN-STN – a typical short-haul flight operated by FR – is -3.9%, while for LH's short-haul FRA-LHR and long-haul FRA-SIN it amounts to –1.7% and –3.7% respectively. Passenger demand for LCCs like FR will be more negatively affected by soaring fuel prices than demand for full service network carriers like LH. The higher demand reduction for FR results from a higher relative fare increase compared to LH as well as a higher share of more price-sensitive leisure travelers. Compared to short-haul routes like FRA-LHR demand for long-haul routes such as FRA-SIN will be more affected due to the relative strong increase in ticket prices and despite lower price elasticities for long-haul travel. Without fuel hedging (α=0), the short-term impact on passenger demand is even stronger. The estimated change in passenger demand due to higher ticket prices for HHN-STN is –16.2%, –6.7% for FRA-LHR and –14.2% for FRA-SIN.

Table 3 does not differentiate the relative price increases with regard to leisure and business market segments. As the average ticket price per passenger is higher for business travelers compared to leisure travelers, the reduction in leisure demand is even stronger and the reduction in business demand lower than shown in Table 3.

This section only estimated the isolated effect of a 50% spot price increase for crude oil transmitted via higher ticket prices on passenger demand. This cause-and-effect chain corresponds to the direct impact of higher fuel prices. There will be indirect impacts of soaring fuel prices, for example, a reduction in

Table 3. Demand reduction due to higher ticket prices

Route (carrier)		HHN-STN (FR)	FRA-LHR (LH)	FRA-SIN (LH)
Avg. price elasticity	Business	−0.7	−0.7	−0.265
	Leisure	−1.52	−1.52	−1.04
Share of business travelers		25%	50%	50%
Rel. price increase	$\alpha = 0.67$	3.0%	1.5%	5.6%
	$\alpha = 0$	12.3%	6.0%	21.7%
Change in demand	$\alpha = 0.67$	−3.9%	−1.7%	−3.7%
	$\alpha = 0$	−16.2%	−6.7%	−14.2%

air travel demand resulting from lower disposable incomes. Looking at the calculated short-term reduction in passenger demand of 16.2% for typical short-haul services operated by LCC and 14.2% for long-haul services in the no fuel hedging scenario, higher ticket prices due to soaring fuel prices strongly influence commercial aviation. However, this direct impact may be compensated by other factors influencing travel demand. Fuel surcharges levied by airlines in recent years did not keep aviation from growing more than 5% annually (Scheelhaase, Grimme & Schaefer 2007). In addition, the estimated changes in passenger demand for services offered by FR and LH have to be set in due proportion to the future growth trend in commercial aviation, especially the currently expected growth rates for European LCCs that go well beyond 5% per annum. The reduction in air travel demand caused by soaring fuel prices may only confine the overall demand increase.

4   Alternatives to Kerosene as Jet Fuel

The previous results show that the rate of air traffic growth, constrained by scarcity of kerosene, will be much lower – or even negative – than unconstrained air traffic growth, especially leading to a strong reduction of demand for leisure traffic and long-haul services. Hence, the entire aviation industry has to look beyond the fuel-efficient '3 liter aircraft' and search for new groundbreaking ways to become less dependent on fossil fuels. This section provides a brief overview of how to save fuel or even replace kerosene as jet fuel.

At present, aircraft and engine manufacturers improve aircraft design (for example blended wing aircraft) and fuel-efficiency of engines in order to reduce fuel consumption. Fuel saving strategies by airlines include shorter air routes, carrying less minimum fuel, increased fuel blending, shorter sector lengths, modern fleet, increased load factors and more efficient ground operations (for example reduction of ground delays). All these efforts contribute to fuel conservation by commercial aviation but do not provide a substitute to conventional petroleum kerosene.

Kerosene is considered the ideal jet fuel. First reason is its high energy content. The energy content of fuel is measured as specific energy which is the energy content per unit mass (joules/kg) and as energy density which is the energy per unit volume (joules/liter). The high energy content of kerosene positively affects the total size and weight of the aircraft. Operationally, the heavier the aircraft is at takeoff, the more fuel is required to lift it into the air. With regard to safety criteria, the Jet A-1 kerosene used in commercial aviation has a high flash point of not lower than 40° Celsius reducing explosion hazards and a low freezing point. Kerosene also does not contain or absorb water which means that in cold temperatures no ice crystals form which block fuel filters and ultimately lead to fuel starvation. These safety over a wide temperature range is an important selection criterion for jet fuels.

Below alternative fuels (synthetic kerosene, bio-fuels and liquid hydrogen) and a new aircraft propulsion technology (fuel cells) are presented and briefly evaluated with reference to Smith (2006), Chevron (2006), Boeing (2007) and Masson & Luongo (2005) along following criteria: high energy content, safety, environmental impact, availability and price. Ethanol and methanol are not considered because of their unfavorable properties at jet fuel.

Synthetic kerosene: This is a carbon-based fuel synthesized by using a Fischer-Tropsch conversion process. According to the raw material used three types of synthetic kerosene are differentiated:

- Biomass to liquid (BTL).
- Gas to liquid (GTL).
- Coal to liquid (CTL).

Today, synthetic kerosene is only approved in commercial aviation as a blend with petroleum kerosene despite of having basically the same energy content and safety qualities. Semi-synthetic fuels (50 percent normal fuel and 50 percent synthetic fuel) for the aviation industry have been produced in South Africa since 1999 (Chevron 2006).

BTL is more environmentally clean than GTL and CTL as the combustion process of BTL releases carbon dioxide (CO_2) in the same quantity as the plants have absorbed from the atmosphere during their growth process. However, the CO_2 benefits of BTL must be assessed by life cycle analyses considering emissions generated by cultivation, processing and transport. BTL can be produced from almost any type of plants and offers new perspectives for farmers but also implies the risk of competition with food production.

Bio-fuels: They refer to fuels derived from feedstock such as rapeseed, soybeans or algae without a Fischer-Tropsch synthesis as in the case of BTL. Bio-fuels have a somewhat lower energy content than kerosene (Chevron 2006). The primary concern with the use of bio-fuels are their low temperature properties with freezing points near 0° Celsius, much higher than the maximum freezing point of petroleum kerosene (–40° Celsius). With additives the low temperature operability at cruising altitudes of bio-fuels can be improved. There are doubts that bio-fuels can be mass-produced affordably because of limited farmland (Boeing 2007). For these reasons, bio-fuels are currently not considered as alternative jet fuels on their own but more suitable for blending with kerosene.

Hydrogen: Liquid hydrogen is the liquid state of the element hydrogen. It is probably the most commonly discussed long-term alternative to kerosene. Hydrogen provides 2.5 times the energy per kg than kerosene but is also about four times more voluminous. Liquid hydrogen is non-corrosive. A major potential advantage of hydrogen compared to kerosene is the significant reduction of harmful emissions. The primary combustion product of hydrogen is water. A negative byproduct of its combustion is water vapor as greenhouse gas. Depending on how hydrogen is produced there are significant CO_2 emissions generated during its life cycle.

Today, hydrogen is expensive to produce and difficult to store. Due to the large volume and the requirement to cool down hydrogen to the liquid state (–253° Celsius), the cryogenic storage of hydrogen constitutes a major challenge for aircraft manufacturers. A hydrogen powered aircraft will look very different from today's kerosene aircraft. Hydrogen will not be stored in conventional wings because of pressurization and insulation requirements. The positioning of fuel tanks in the fuselage results in an enlarged fuselage or less passenger capacity. Ensuring explosion safety of cryogenic aircraft is a challenge. Hydrogen will also require a radical change in engine design. Yet the Russian aircraft manufacturer Tupolev managed these technical challenges with the cryogenic fuel aircraft TU-155 performing its maiden flight already in 1988 (Tupolev 2007).

Hydrogen aircraft also pose a major challenge for airport infrastructure which at present is only designed for kerosene aircraft. A prerequisite for a change from kerosene to hydrogen already in the transition stage is the global availability of two parallel fueling systems at airports. Hence, a transition to hydrogen-powered aviation may take decades, especially considering the long life-span of aircrafts currently in operation.

Fuel cells: They have been used in spacecrafts since the 1960's to power auxiliary engines. Experimental aircraft powered only by a fuel cell supported by lightweight batteries during takeoff and climb is on its way. A fuel cell is an electrochemical device that converts hydrogen directly into electricity and heat without combustion. Fuel cells are emission-free and quieter than hydrocarbon fuel-powered engines. The main challenge is to develop compact and lightweight electric propulsion systems with more power. Today, using fuel cell technology as primary power for a passenger airplane leads to a propulsion system several times heavier than conventional aircraft engines and still far from their efficiency. However, chilled superconducting magnets carrying electricity without resistance have been proposed that may allow for lightweight and powerful electric jet engines in the long run (Masson & Luongo 2005).

Table 4. Assessment of alternative jet fuels and fuel cells relative to petroleum kerosene

Criterion		Energy Content	Safety	Environmental impact	Availability and price
Synthetic kerosene	BTL	o	o	+	–
	CTL	o	o	–	–
	GTL	o	o	–	–
Bio-fuels		–	–	+	?
Hydrogen		+	?	+	–
Fuel Cells		–	?	+	–

Table 4 summarizes the pros and cons of jet fuel alternatives relative to petroleum kerosene along selected criteria. "+" indicates that the potential substitute performs better with regard to the respective criterion, „o" suggests equal and "-" worse properties compared to conventional kerosene. The assessment does not account for ground-breaking technology developments and, hence, has to be regarded as preliminary. Evaluating alternatives to petroleum kerosene in the near future, synthetic kerosene holds the greatest promise as it basically can be used in existing aircraft either alone or blended with petroleum kerosene. The main problem for synthetic kerosene with the exception of BTL is the large amount of CO_2 generated during production. In the long run, hydrogen seems to be a promising candidate to replace kerosene if safety standards of civil aviation can be secured but asks for a fundamental change in aircraft design and new ground infrastructure at airports.

5   Conclusions

Conventional wisdom in commercial aviation is that global air traffic will continue to grow in the coming decades. This implicitly assumes no constraint in traffic growth due to finite oil resources. This is in stark contrast to studies that estimate peak oil sometime between now and 2040.

This paper analyzed the short-term economic impact of soaring fuel prices on commercial aviation. The time horizon was only one year allowing fuel hedging by airlines to balance increasing spot prices. The analysis was restricted to the direct effect of higher kerosene prices on operating costs, fare levels and passenger demand. Indirect effects on passenger demand resulting from a reduction of the purchasing power, an increase in unemployment and higher costs for other input factors besides kerosene were not considered. The analysis also ignored possible political crisis and economic shocks for oil importing countries forced to spend significantly more on their energy purchases. Hence, the scope of this paper has been somehow limited.

However, the limited approach already shows that the rate of air traffic growth constrained by scarcity of kerosene will be much lower – and may even be negative – than unconstrained air traffic growth, especially with regard to price-sensitive leisure demand. Services offered by low-cost carriers and long-haul services will be most adversely affected by higher fuel prices. Further, the impact of soaring fuel prices largely exceeds the impact of the proposed EU emission trading system (ETS) for the aviation industry. This leads to the question whether ETS is actually needed in view of finite supplies of fossil fuels that may restrict or even terminate air traffic growth. In addition, high fuel prices are a strong incentive to use more fuel-efficient engines, to optimize minimum fuel policies, to improve air routes and ground operations, etc., in the same direction as intended by ETS.

The fuel price development will also influence the typical air service pattern, for example, there may be a renaissance of technical stops for re-fueling on intercontinental routes or more point-to-point traffic in order to avoid fuel burning detours via hubs. To avoid high fuel costs, regional carriers have already replaced regional jets on some routes by turboprops. The in-depth analysis of the relative economic benefit of competing services patterns and the use of turboprops instead of regional jets in times of high fuel prices is an interesting issue for further research.

Peak oil will happen, the open question is when. It is a problem of particular importance for commercial aviation as jets are not as fuel-flexible as ground vehicles. The long-term implications of finite oil resources go beyond future air frame designs or increased blending of other fuels with kerosene. More research than today should be devoted to the economic evaluation of kerosene substitutes in combination with the associated future requirements for airline fleets and airport infrastructure.

References

Boeing (2007): Alternate Fuels for use in Commercial Aircraft, Seattle.

Chevron (2006): Alternative jet fuels, supplement to Chevron's Aviation Fuels Technical Review, Houston.

Deutsche Lufthansa AG (2007), Balance 2007, Cologne.

Energy Information Association (EIA) (2008): International Energy Outlook, Washington.

Gillen, D., Morrison, W. and C. Stewart (2004): 'Air travel demand elasticities – concepts, issues and measurement', study commissioned by the Canadian Department of Finance, Ottawa.

Government Accountability Office (GAO) (2007): 'Uncertainty about future oil supply makes it important to develop a strategy for addressing a peak and decline in oil production', report to congressional requesters, Washington.

International Air Transport Association (IATA) (2008): Financial Forecast, Montreal, June 2008.

Masson, P.J., and C. A. Luongo (2005): 'High power density superconducting motor for all-electric aircraft propulsion', IEEE Transactions on Applied Superconductivity, 15, 2226-2229.

Morrell, P., and W. Swan (2006): 'Airline jet fuel hedging: theory and practice', Transport Reviews, 26, 713-730.

Scheelhaase, J., and W. Grimme (2007): 'Emissions trading for international aviation – an estimation of the economic impact on selected European airlines', Journal of Air Transport Management, 13, 253-263.

Scheelhaase, J., Grimme, W., and M. Schaefer (2007): 'European Commission plans emissions trading for aviation industry', Aerlines, e-zine edition, 36, 1-5.

Smith C. (2006): 'Aviation and oil depletion', presentation at Energy Institute, London, November 2006.

Tupolev (2007): 'Crygenic aircraft', www.tupolev.ru/English/Show.asp?SectionID=82 (visited Nov. 20, 2007).

The Future of the Passenger Process

Resolving the Bottlenecks at the Check-in and Security Checkpoints

Patrick S. Merten

1 Introduction

The passenger process at airports is undergoing an overhaul: mobile tickets, check-in kiosks, web and mobile check-in, as well as electronic and mobile boarding passes are transforming the traditional process. This article addresses the *future of the passenger process at airports* and examines the changes from a passenger's point of view as well as from the perspective of airline and airport managers. *Optimising passenger flow* is the general objective of both groups. This specifically translates into *resolving the bottleneck at the check-in area and at security checkpoints*. These two issues are discussed below in detail. The insights presented below were gained from an expert study as well as a passenger study based on a survey.

2 Details of the Studies

The following discussion is mainly based on two surveys, conducted in 2007 and 2008.

2.1 Passenger Study

The passenger study involved a survey of around 1,400 passengers at Berlin airports. The survey was aimed at analysing *acceptance of technological innovations in the passenger process*, in particular:

- mobile ticketing,
- kiosk, web and mobile check-in,
- electronic and mobile boarding passes,
- mobile information services.

The survey further addressed passenger attitudes to a *city check-in* and *alternative baggage drop-off facilities* like at central urban locations as well as baggage pick-up services from homes and offices.

Special attention was paid to border control and security checks. Detailed responses were gathered with regard to passenger attitudes towards *RFID chips and biometric data in passports*. The study analysed passenger perception of biometric controls also at security checkpoints with special focus on the *acceptance of full body scanners*. Furthermore, current practices for *checking passengers' carry-on baggage* were subjected to a critical review.

2.2   Expert Study of Airline and Airport Managers

Apart from the insights gained from the passenger study, the present article is also based on insights from a *Europe-wide survey of airline and airport managers*. In-depth expert interviews were conducted over a period of one year with managers from 20 percent of all European airlines and airports which handle over a million passengers per year. The interviews touched upon all aspects of the passenger process as in the passenger study – the same aspects were discussed for the study but from the perspective of airline and airport operators. The goal of the study was to generate *trends and scenarios for a future passenger process* and identify how probable and desirable they are based on current developments.

3   The Future of Ticketing, Check-in and Boarding Passes

In recent years, airlines have mainly focused on the distribution phase of the passenger process. Key indicators of this development were the revamping of the Global Distribution Systems (GDS) and evolution of alternative distribution systems. The introduction of e-tickets rode the wave of successful distribution via the Internet. According to IATA, the aviation industry achieved the target of nearly 100% e-ticketing in 2008. E-tickets are electronic documents with a 1D or 2D barcode which passengers can print themselves and present as proof of booking at the time of check-in. In practice, passengers increasingly need to only present a valid proof of identity at check-in time, as a result of which the e-ticket is often nothing more than a receipt. Concurrently, mobile ticketing is also seen to become increasingly common: the m-ticket is an electronic image of a 2D barcode and some further information which is sent to a passenger's mobile phone and must be presented at the time of check-in. A pertinent question here is why passengers should need a m-ticket to check in when even the e-ticket has become obsolete. Following this line of reasoning, some airlines have already decided against m-ticketing. Irrespective of these trends, passengers can make a booking by accessing the Internet via their mobile phone and receive an e-mail and e-ticket as confirmation. Based on these general contours of the distribution phase, experts discussed the check-in phase as follows.

3.1   Changes to the Check-in Procedure

After successfully renewing the distribution environment, airline and airport managers are now shifting their focus to passenger check-in. Based on the *trend towards Internet distribution and the success of e-tickets,* self-service kiosks were progressively introduced in airports in recent years. However, status quo analysis shows that only around *40 percent of all airlines and airports in Europe use the so-called CUSS (Common User Self Service) kiosks.* In contrast, *65 percent of all European airlines already offer web check-in platforms*, although this variant was implemented only two years ago. Expert interviews show that a fundamental distinction can be made between two competing trends: One group of experts believes that over 50 percent of all check-ins in future will be done at airports with passengers increasingly using the check-in kiosks. In contrast, the other group of experts expects that 70 percent of all passengers will check-in off-airport, i.e. using web or mobile check-in facilities. The target for these projections is set for 2015.

For airlines, the choice between the counter and self-service check-in alternative depends, among other reasons, on the *passenger and service portfolio.* As such, projections expect that the number of check-in counters will be reduced in future but the facility will be retained for *exception handling* i.e. to cater to special requests and passengers who may need assistance. However, expert opinions are sharply divided over the future of check-in kiosks. The first group will stick to the *investments in kiosks* while the second group of experts thinks that the era of kiosks has already come to an end.

In contrast, the choice of check-in variant for airports depends on *infrastructural conditions.* As airports increasingly generate *revenue from non-aviation services*, the web and mobile check-in variants help to free up valuable space in the check-in area for service offerings. Airport experts also point out that shifting passengers from the counter to kiosk will not resolve the check-in bottleneck because, in the worst case, *queues will form at both check-in alternatives at the airport.* On top of this, *additional counters have to be set up for separate baggage handling.*

Irrespective of which of the three self-service check-in alternatives passengers prefer, the traditional link between *baggage check-in and passenger check-in will be severed.*

3.2   Decentralised Baggage Handling

The separation of check-in and baggage handling process indicate a wide range of trends at different airports. They range from *baggage drop-off at centralised city offices* to *baggage pick-up service from homes and offices.* The passenger survey revealed a *rate of acceptance of approximately 50 percent.* These passengers were mostly also prepared to pay for the service as shown in Figures 1 and 2. The *airport drop-off alternative* will continue to exist as this variant gained acceptance since check-in kiosks were introduced. Experts also proposed *relocating baggage drop-off to the car park or to the bus and train arrival point at the airport* which would free up space at the check-in area for other service offerings.

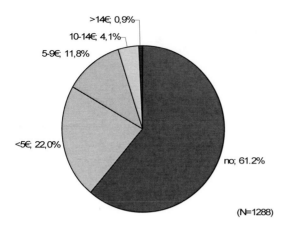

Fig. 1. Willingness to pay for a centralised baggage drop-off facility in the city

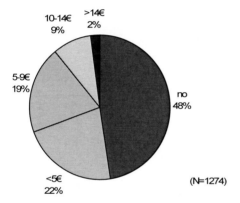

Fig. 2. Willingness to pay for a baggage pick-up from home or office

The following sections discuss the three alternatives to the traditional check-in counter from the perspective of passengers.

3.3   Acceptance of Check-in Kiosks

Check-in kiosks have gained acceptance in the meanwhile after a relatively pro-longed start-up period. The survey of around 1,400 passengers indicated that 82 percent of passengers have heard of this check-in facility and 73 percent appreci-ates the facility. Following initial negative reports, *positive usage experience is currently at 80* percent.

Nevertheless, the survey revealed that the *check-in counter is still clearly the first preference across all age groups,* followed by the kiosk and web check-in variants. Figure 3 shows ranking position one among the different check-in alter

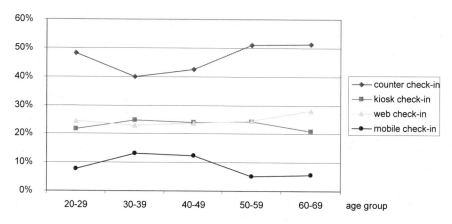

Fig. 3. Ranking position one of check-in alternatives

natives. Note should be taken of passengers in the 30-49 age group. Most respondents of this group ranked the counter either as the first or last choice. Kiosk and web check-in appear to be the alternatives for this group. While Figure 3 indicates that *kiosk and web check-in are cited as the first priority by equal number of passengers (over 20 percent for each)*, a further analysis showed that *kiosks are currently the second preference*, while *web check-in ranks third*.

3.4   Web and Mobile Check-in Perspectives

Web check-in is effectively offered by a growing number of airlines for only about the last two years. Although it has been on the market for a short time, the study reveals *an almost equally positive measure of usage experience*. However, the number of passengers who have heard of the web check-in option is 20per cent lower. The development is even more impressive if the development of web check-in is compared with that of the kiosk which had to deal with prolonged start-up problems. Figure 4 compares passenger attitudes towards usage and usage experience of all three check-in alternatives.

It is obvious that very few passengers have yet heard of mobile check-in or have used the alternative. It is unexpected, however, that even *perception and attitude towards usage of mobile check-in* is negative. Further, the *assessment of technophile early adopters is also relatively negative*.

As illustrated, *web check-in is markedly higher for young professionals (age group 30-39)*. It may be assumed that this fact will most certainly continue to have a positive impact on web check-in in the coming years especially when the current generation of web-oriented passengers in the 20-29 age group moves into the next age group. At first glance, it is unexpected that *web check-in is clearly preferred to kiosk check-in among passengers in the 60-69 age group*. This phenomenon can be explained by the fact that passengers in this age group and older may prefer to

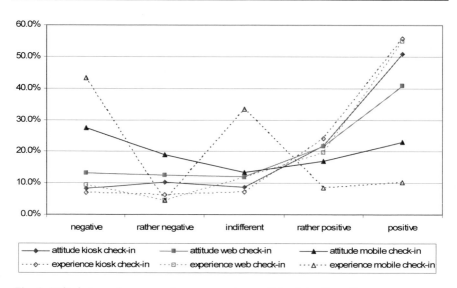

Fig. 4. Attitude towards usage and usage experience of check-in alternatives

check-in time and place independently at home, without time pressure but with assistance of their relatives, rather than *standing in an airport queue with baggage under time pressure without the assistance at the kiosk check-in*.

In order to generate more precise recommendations for action for the airlines and airports, multivariate analyses were conducted to determine which passenger characteristics have a significant impact on the acceptance of the new check-in

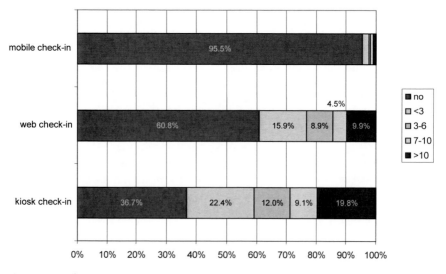

Fig. 5. Usage frequency of check-in alternatives

alternatives. The findings unexpectedly reveal that neither age nor gender has a constant, significant or strong influence on the acceptance. On the other hand, frequency of flying is clearly seen to have a positive effect, more in the case of business travellers than in the case of private travellers. It is therefore not surprising that membership in a frequent flyer programme is a significant criterion for the acceptance of technological innovations in the passenger process.

3.5 Convergence of the Distribution and Check-in Phases with Electronic and Mobile Boarding Passes

Interviews with experts also examined a scenario which combines booking and check-in under the premise that in future both processes will be handled either via the Internet or the mobile phone. In this scenario, passengers can *already check in at the time of booking* and will directly receive a boarding pass instead of an e-ticket. This is possible as according to IATA, airlines must switch from ATB2-compliant boarding passes to a *2D barcode-based standard* by 2010. PDF417 is the applicable standard for *e-boarding passes* while the IATA plan currently includes three alternatives for *m-boarding passes*, i.e. Aztec, Datamatrix and QR. This development will be primarily *driven by the increase in web and mobile check-in* which need these types of boarding passes.

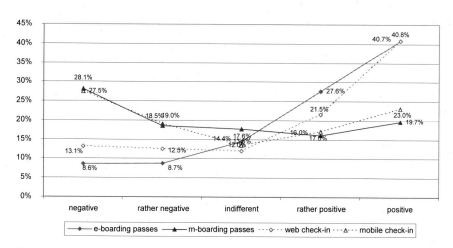

Fig. 6. Comparison of attitudes to e- and m-boarding passes for check-in

4 Technological Innovations at Borders and Security Checkpoints

The following sections deal with different technological innovations which could potentially relieve the bottleneck at border control and security checks.

4.1   Identification Authentication Checks

The surveyed passengers were first asked at which points they considered identification and authentication checks necessary. The study showed that *around 75 percent of respondents approve of a control during check-in and at the security checkpoint* while *only 57 percent considered a check necessary at the boarding gate*. The passengers were then asked for their preferred methods of identification. Despite current developments, *no more than 13 percent of the passengers expected any of the new identification alternatives such as frequent flyer card, credit card or mobile phone.* The study unambiguously showed that the *identification card and passport are the only identification documents* with high acceptance, in this case almost a 100 percent. On the other hand, a few passengers even cited biometric checks as the desired alternative. This aspect is treated in greater detail in a similar context below.

4.2   Embedded RFID and Biometric Chips in Passport

Regarding embedded RFID chips and chips with biometric data in passports, the study showed that *44.7 percent of the passengers accepted the RFID chip solution* while only *39.5 percent approved of storing biometric data.* In view of the political dimension of this topic and public discussion of the topic in the media, these figures should rather be considered positive and high.

The two technologies mentioned above were also evaluated with regard to the individual dimensions of passenger acceptance as shown in Figure 7.

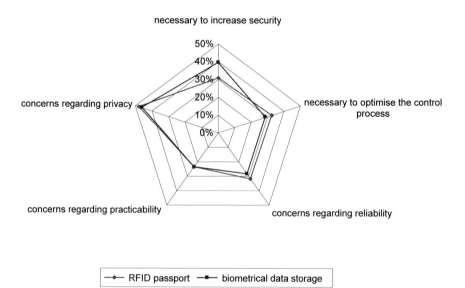

Fig. 7. Acceptance of embedded RFID chips and biometric data in passports

The results clearly show that the *necessity to store biometric data to increase security* is perceived to be much higher than the admitted necessity for process optimisation, in the view of passengers. On the other hand, concerns regarding the practicability and reliability of the two technologies ranked lower while *serious doubts with regard to data protection and privacy* were expressed.

Finally, passengers were also asked about various kinds of biometric checks. The *approval rating of fingerprint scanners amounted to 57.3 percent and that of iris and facial scans to 45 percent*. Alternative methods such as hand-palm geometry, movement and voice analyses scored marginally in contrast.

4.3   Passenger and Carry-on Baggage Checks

In addition to the border controls, the study also assessed passenger and carry-on baggage security checks. As Figure 8 shows, *positive assessment of checks falls with every additional step in the process*. Still, overall assessment of checks is very positive except for the recent regulation on carrying fluids on board.

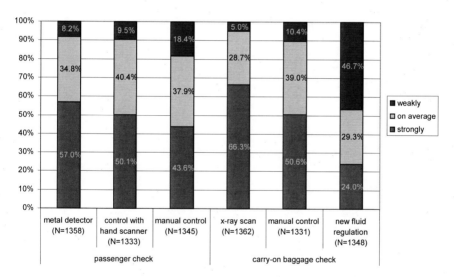

Fig. 8. Acceptance of passenger and carry-on baggage checks

In addition to the analysis of the current process and the deployed technical solutions, *full body scanners* were also evaluated in detail. A positive view of the solution in principle was taken by 34.4 percent of the surveyed passengers, while 19.9 percent are indifferent but 45.7 percent tend to reject the solution. Figure 9 shows in detail how full-body scanners were assessed in regard to individual aspects of acceptance. As with the technologies discussed earlier, the results clearly show that concerns over reliability and practicability were evaluated far lower than *concerns over data protection and health aspects in this case*.

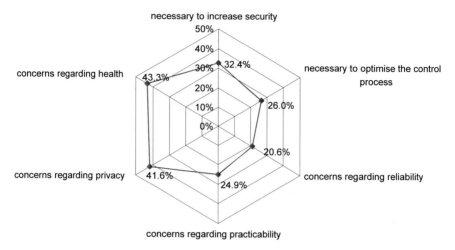

Fig. 9. Acceptance of full body scanners

4.4   Opinion of Experts

Despite the results from the passenger survey, implementation of new border control and security check facilities is independent of passenger opinion even though passenger acceptance will help. Concerning the changes in the security checks, airports primarily find themselves confronted with non-existent or inadequate national as well as international guidelines and laws. Furthermore, airports are mostly only infrastructure providers for these processes. Thus, issues such as the full body scanner are critical for airports in view of high investment costs and additional space overheads. The IATA Simplifying Passenger Travel (SPT) initiative is dealing with the introduction of a biometrical line and a priority line and should thus be watched closely.

An alternative suggestion from the experts to optimise processes based on existing infrastructure proposes to de-couple and *separately execute all time-consuming process steps* such as placing all external apparel, carry-on baggage and contents e.g. notebooks and liquids in separate containers prior to scanning. Such a step would reduce congestions and could speed up passenger flow, the experts add.

5   Summary

The selected results of the passenger survey and expert study presented above clearly demonstrate that the passenger process is undergoing some fundamental changes. The opinion of experts and findings on the passenger acceptance can be interpreted further to conclude that classical process steps in their present form will no longer exist in the future. The more pertinent question in this context is whether multiple alternative processes will exist concurrently in future or whether processes will be consolidated whereby some, even newer, process alternatives will diminish. For further information on the topic, the passenger survey and expert study as well as potential scenarios for the future of the passenger process, please contact patrick.merten@aviation-research.de.

6 Transcript of Statement by Martin Gaebges, General Secretary, Board of Airline Representatives in Germany (BARIG)

Let us start with a small scenario and imagine the following business trip: you arrive at the airport with your suitcase and briefcase at the airport about half an hour prior to flight departure. You proceed with your ticket through the security check, walk straight to the departure gate where a member of the service staff hands you a boarding card. If you have any baggage that needs to be checked-in, you hand it over so that it will be directly handed over to you when you arrive at the destination. If you still have time on your hands, you could briefly walk over to the duty free shop and buy an attractively priced bottle of fine cognac and stick the bottle into your carry-on baggage. The flight is called and you proceed to board the aircraft and a few minutes later you are up in the air. Does all this sound like science fiction, a vision of the future?

This is much rather a nostalgic memory of past times, for this is what air travel was like 30 years ago. Seniors among you, those of about my age, will remember reporting times of about 20 minutes prior to departure. Several airlines had reporting times on selected routes of even 15 minutes prior to departure. What reporting times and air travel procedures are today, I need not describe to you in any detail – many among you have probably had to survive the ordeal in recent days.

The great paradox here is that our understanding of passengers and passenger processes has steadily increased over the decades. Airline customers check in from home or via the Internet, they carry their own baggage and deposit it at the automatic check-in machines. They fill in additional forms and facilitate security checks by visibly packing their toothpaste in transparent bags, pull out and show their laptops and pack them in again etc. The interesting point is that customers are made to do more and put up with less comfort and still, ground operations of air travel are becoming ever slower.

On many short routes, air travel has lost its benefits compared with other modes of transport.

It is certainly in the interest of customers as well as in the interest of competition – the airlines at the end of the day – that passenger processes are kept as economical as possible. There is no question, of course, that security takes topmost priority.

Still, we must critically review passenger processes repeatedly in regard to the following aspects: the real objective of passengers is to travel quickly from A to B, quickly and as comfortably as possible. Passenger processes at the airport are therefore services so we must ask whether some links in the service chain are at the risk of becoming serf-serving. On top, the government bears the responsibility for the safety of citizens including those of travellers from all over the world who are entering or leaving the country. Are the government agencies involved really making use of all available and necessary means to make security as efficient as quick as possible? Which security measures actually represent a real increase in security and which are more suggestive of demonstrative action to serve political ends? Do innovative technology and automated processes really speed up travel or are they in reality means to pass on costs to passengers? How will air travel develop to remain competitive with/despite passenger handling compared with other modes of transport? All these are questions which we shall be able to address in greater detail in the course of discussions.

7 Transcript of Statement by A.S. Viswanathan, Head of Transportation/Airports, Siemens IT Solutions and Services, Mumbai/India

Each of us is not only a designer and a technologist helping to improve the systems, but each of us is also a consumer of the airlines industry.

Take the fact that I'm here on time to talk to you: I came by a flight which just left Frankfurt at 9 o'clock and it was here by 10.30. So in regard to what people are talking about here, today's processes in the airlines are highly optimized considering the existing infrastructure. Having said that, to exploit productivity and increase innovations requires a lot of technology.

And this is an area where technology is used as my friend Patrick, has clearly illustrated. Technology is not a substitute for a process. The simplification of passenger travel and the bottlenecks is continuously different, changing. What I would say about the Greenfield airports which are coming in Asia, is that there is a chance for each of us to break free from the legacy. Start and build up a system which will have an integrated view of the entire process. That is what has happened in the Greenfield airports in India. We are maintaining it and I'm still counting on my fingers to see that whatever I speak about hopefully comes true in a foreseeable amount of time.

But the point I want to make is that the entire architecture has to facilitate an integration of technologies, passenger-centric technologies and environment-centric technologies.

Another aspect of Greenfield airports is that we are finding very significant distinctions, differences in terms of an approach because these are no longer airline-centred operations.

Each of the airports wants to attract passengers and airlines to the hubs and this is changing the whole dimension of who is running the whole show. It is not longer airline driven, it is airport driven.

And the second aspect is the commercial exploitation of the airport because whenever an airport comes to a city, you are also talking about an equal system that is generating the economy. For the airport operator, that means non-aviation income, so whenever we are designing the IT systems, we have seen that people are more interested in seeing the commercial dimension of the airports – how quickly can you do the building and can we monitor the assets for the whole airports, which unfortunately is not the case today. As we have seen, the other aspects, e.g. security processes so on, I should confess, we are in a fairly primitive state, even at the Greenfield airports which we have built, unfortunately. We are confronted with a choice, that is, technology vs. individual freedom. It is not an introduction of technology by itself, but the regulations that really dictate the use of technology.

I believe that to solve one of the biggest bottlenecks, which can really help us realize the dream of the speed, comfort and security, we need all the players including also the government authorities, the federal authorities, to sit across the table and talk about the processes.

Otherwise, we would really not draw the benefit of new systems, new processes, not even in Greenfield airports.

8   Transcript of Statement by Rainer Schwarz, Spokesman for the Managing Board, Berlin Airports

I would like to say this right at the outset, despite all our faith in modern technology there is something that we should not forget. Even in the passenger handling processes of airports and in airlines, we are dealing with human beings and this is an important factor that we should keep in mind in all our discussions. Here in Berlin, we are presented with the unique opportunity not only to design a brand new airport but also to build it so that it is commissioned in November 2011. We are presented with the unique opportunity to also introduce new processes in the new airport, processes that existing airports can often introduce and optimise only with a great deal of difficulty.

We are putting in a great deal of thought into the matter, you can see that every study that we commission deals not only with what technology can achieve but also with customer acceptance and emerging trends. Excellent simulation tools are available these days where we can see in the process chain that you have just presented, how each links fits with the next. Of course we are very keen to optimise on space and costs at the end of the day. We run these things individually through the simulation and in actual fact we see that there is a trend in the airline industry to pass on tasks to the customer. The more low-cost players budge in, the more tasks are passed on to the customer – tasks that were earlier handled by the airlines – the underlying expectation being that customers should do most tasks from home or anywhere, so long as they don't bother the airlines.

For us as airport operators this is a pretty nice development at first glance because we only have a limited amount of space at our disposal and as you have just said, we do of course have a keen interest to see that this space is given to those who pay most for it. So there is competition for space even in our industry. So any space that we do not need for check-in or security facilities is space that we obviously want to use for duty-free or other non-aviation revenue sources.

Still, our principle is to tailor our offerings on the lines of the cafeteria-concept, i.e. we ask our customers, the airlines, what they want. The low-cost airlines who want to move everything outside need relatively little space and so they will have different accesses. We offer full-service airlines the option of offering passenger handling processes as in the sixties also at BBI. Of course this needs space and space costs money. We monitor the situation very carefully and we charge for services. We want to make sure that we do not see a situation like supposedly handicapped people who were used to getting the service free but when we charge them €10 for special care, we often see at the airports that they throw away their walking aids and walk perfectly normally. So we need to view these things in a certain context. I believe our task as airport operators is to offer a choice to our airline customers on what they would like to have. To this extent we'll only be very glad to resort to technical aids. I believe this approach will present us all the opportunity to integrate state-of-the-art technology. In our assumptions so far we have already factored in some percentage figures with regard to staffed interactions (e.g. check-in) and automated operations. We should certainly be much clearer in our thinking, look ahead and anticipate future development.

One aspect has been completely disregarded in our discussions and I'd still like to bring it up here. Among all the issues about which technology is really necessary and what still needs to be done, many process decisions also depend on the market situation. BBI, the airport that we are building is scaled for 20-40 million passengers. We already have about 20 million passengers at Berlin Airports so the new airport can handle double the capacity. So BBI can handle 40 million passengers and that is a very good segment because all airports that are larger have very many transit routes. That makes the whole thing so complex that passengers will find it formidable. Take Heathrow or Frankfurt for example, you cannot safely negotiate these airports without bringing a spare shirt along! What I'm trying to say is that the market situation is critical: irrespective of what technology we have, how we position ourselves as an airport, whether we can offer non-stop connections so that I do not have to spend a whole lot of time in synchronizing extremely complex connections – all these issues will also simplify passenger handling processes.

I believe, we all have a great opportunity here in Berlin as we are building a Greenfield airport, and can integrate all these processes right from the beginning. We have the opportunity to offer at the end of the day, what passengers need and want. I also believe that the scale is necessary because we will be one of the few airports in Germany which will be in a position of offer capacity growth potential at the infrastructure level.

Travel Technology

The PhoCusWright Consumer Technology Survey Second Edition

Cathy Schetzina

This chapter studies a range of consumer technologies relevant to travel and assesses U.S. online traveler familiarity and usage patterns (section 1), as well as the degree to which these technologies influence purchasing behaviour (section 2). Moreover the survey gives insights into the influence of consumer technology on travel (section 3). The nature of social media – especially traveler reviews and social networks – is highly personal. Travelers are interested only in the information or promotions that are most relevant to them i.e. reviews from friends and family. So section 4 deals with the social media usage, influence and marketing preferences. Innovative mobile handsets – and an increasing number of available mobile Web sites and applications – gives a range of opportunities. But how will travelers use this? Information to this and other related questions gives section 5 "Mobile Device, Activities and Interest".

1 Familiarity and Usage Patterns

Consumer familiarity with various Web 2.0 technologies may not be increasing as rapidly as previously anticipated, with a recent survey indicating that recognition of most technologies is flat compared to 2007. The sole exception is social networks. Online travelers are becoming more familiar with social networks, although a slightly smaller number report participating in them.

The PhoCusWright Consumer Technology Survey Second Edition studies a range of consumer technologies relevant to travel and assesses U.S. online traveler familiarity and usage patterns, as well as the degree to which these technologies influence purchasing behaviour (see Methodology below).

Online maps and social networks remain the most familiar technologies; 68% of U.S. online travelers reported being very familiar with online maps and 49% reported being very familiar with social networks (see Figure 1). In addition, 40% of online travelers said they were very familiar with 3D maps and virtual tours. Conversely, just 10% said they were very familiar with virtual worlds, while 62%

said they were not familiar with them at all. The increasing ubiquity of 3D maps and virtual tours could give rise to a convergence between mapping and virtual worlds, accelerating adoption through continued growth of familiar technologies.

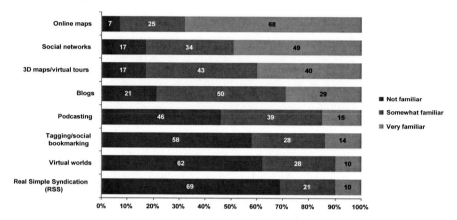

Note: Q7. Please indicate your degree of familiarity, if any, with the following online technologies using a 3-point scale, where 1 = "Not familiar" and 3 = "Very familiar"

Fig. 1. Degree of familiarity with online technologies

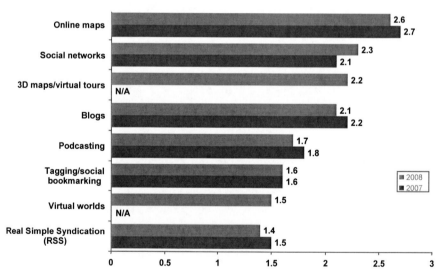

Note: Q7. Please indicate your degree of familiarity, if any, with the following online technologies using a 3-point scale, where 1 = "Not familiar" and 3 = "Very familiar" (means) (2007: males are weighted 2.95 to 1)

Fig. 2. Mean familiarity with online technologies

1.1   Social Networks Gain Ground

Familiarity with online technologies tracks closely to that reported in 2007 (see Figure 2). Online maps, blogs, podcasting, tagging and RSS all yielded similar results. In most cases, online travelers reported being slightly less familiar with the technology. Surprisingly, only social networks gained ground among consumers in the past year, with mean familiarity increasingly to 2.3 in 2008, versus just 2.1 in 2007.

Despite this increased familiarity, a slightly smaller number of online travelers reported joining or participating in a social network. In 2008, 64% reported having joined or participated in a social network, versus 68% in 2007. Travel-related participation in a social network was also nearly even, with 20% reporting ever having done so in 2008, versus 22% in 2007 (see Figure 3). This discrepancy in familiarity versus participation may be due to the increasing media spotlight on social networks during the past year, which could potentially lead to an increase in new participants over time.

There continues to be a much larger number of online travelers who consume Travel 2.0 content compared to those who produce it. Seventeen percent of online travelers reported keeping a travel-related blog in 2008, while 42% read one. Forty percent have posted a travel review, versus 84% who have read one.

Despite this discrepancy, there are some signs that consumers are taking a more active role online (see Figure 4). A larger percentage of online travelers reported posting photos of any kind, including travelrelated photos. Sixty-one percent reported posting photos of any kind (versus 59% in 2007), while 37% have posted travel-related photos (versus 31% in 2007).

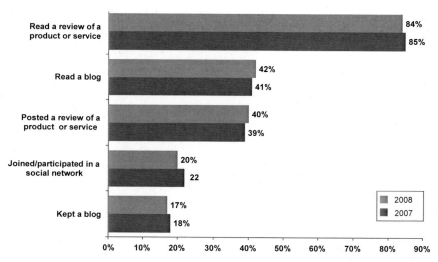

Note: Q8A(ii). Which of the following activities, if any, have you done when related to travel? *Check all that apply.*

Fig. 3. Using online technology for travel-related activities

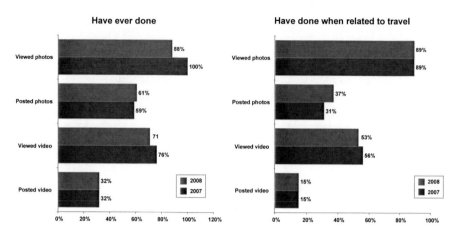

Note: Q10. Which of the following activities have you ever done online and/or done specifically related to travel? *Check all that apply for both columns.*

Fig. 4. Using rich media online

1.2   Online Maps Lead Web 2.0 Use

The use of most Web 2.0 technologies for travel remains far below overall use, with the exception of online maps (see Figure 5). Forty-three percent of online travelers reported watching a video podcast, but only 15% have done so for travel. Thirty-three percent said they participated in tagging/social bookmarking, while just 15% have done travelrelated tagging.

Notably, nearly four in ten online travelers have created an avatar, which is a graphical representation of a user. Avatars have become increasingly prevalent in chat programs, virtual worlds and games. Their potential applications for travel include being used in 3D maps to create a more personalized, immersive environment or as virtual online travel agents or customer service representatives, which could decrease call center costs. To date, however, only 8% of online travelers have created avatars for travel-related activities.

In addition, there was no significant increase in the adoption of Web 2.0 technologies among online travelers in 2008. A handful of activities, including online maps, travel reviews and digital photos, continued to dominate traveler activities. Not surprisingly, these are the technologies that have been most broadly incorporated into travel Web sites. Very few Travel 2.0-specific sites have gained traction. Recent data analysis by Web analytics firm Hitwise revealed only two travel-related social networks within the top 200 travel Web sites based on February 2008 traffic. Despite the sluggish growth for travel social networks, familiarity with social networks in general increased.

The most popular consumer technologies appear to be clustered around rich media, which enables travelers to preview a destination or area, and advice, such as travel reviews. Social networks, 3D maps and virtual worlds represent the newest, most innovative technologies that can fulfill those traveler needs. While familiarity and usage of Web 2.0 and Travel 2.0 tools appear stagnant year over year, these highly travel-relevant technologies seem most likely to attract interest in the future.

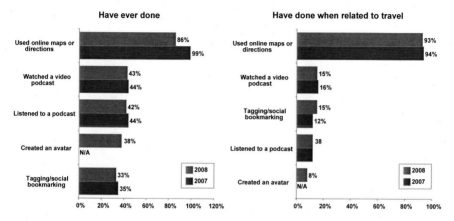

Note: Q11. Which of the following activities have you ever done online and/or done specifically related to travel? *Check all that apply for both columns.*

Fig. 5. Using Web 2.0 technology

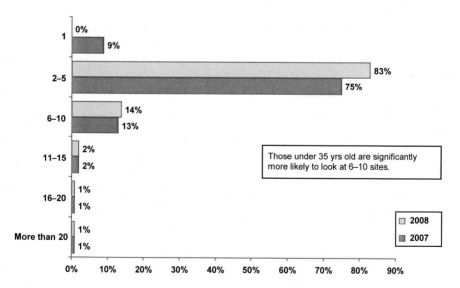

Note: Q2. When planning a trip online, how many Web sites do you usually visit? *Select one.*

Fig. 6. Number of Web sites visited when planning travel online

2   Online Traveler Shopping Behavior

Any travel company aiming to make its travel Web site a one-stop shop, take note: online travelers are all but guaranteed to visit multiple sites. In a recent survey, 100% of online travelers reported visiting more than one site, up from 91% in 2007. The most popular type of site for both beginning a travel search and booking travel is online travel agencies.

Most travelers (83%) visit between two and five sites when planning a trip, up from 75% in 2007 (see Figure 6). The number of travelers visiting between six and 10 sites remains nearly even year over year, and 14% of travelers in 2008 fell into that category. Travelers under the age of 35 are most likely to visit six to 10 sites; 19% of travelers in the 26-34 age group reported doing so.

Comparing prices remains the most commonly cited reason for visiting multiple Web sites; 96% of travelers reported shopping around for that purpose (see Figure 7). The next most common reason is to purchase tickets to events or activities (50%), followed by researching possible destinations (48%). Thirtyfive percent of travelers visit multiple sites to read traveler reviews; this is down slightly from 2007, perhaps because traveler reviews have been integrated into many (if not most) travel Web sites.

Twenty-nine percent of travelers visit multiple sites to view rich media, a purpose that was not analyzed in 2007. Rich media, however, continues to be cited

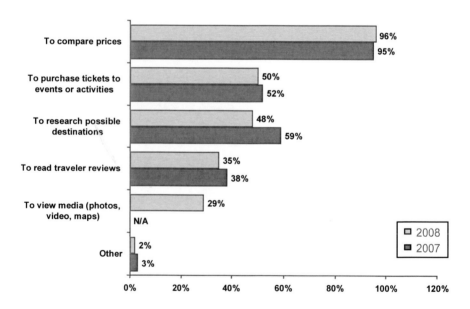

Note: Q2a. Why do you visit multiple Web sites when planning travel? *Check all that apply.*

Fig. 7. Reasons for visiting multiple Web sites

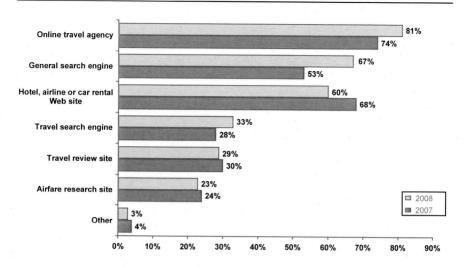

Note: Q3. Which of the following types of Web sites do you usually visit in the course of planning a leisure trip?

Fig. 8. Web sites usually visited when planning travel

as highly influential for travelers making purchasing decisions. Notably, travelers in the 45-54 age group are most likely to visit multiple sites to view rich media; 35% of travelers in this group reported doing so.

Online travel agencies are the most popular type of site to visit in the course of planning a leisure trip; 81% of online travelers do so, up from 74% in 2007.

Supplier Web sites, in contrast, have lost ground – 60% of online travelers visited hotel, airline or car rental Web sites in 2008, versus 68% in 2007.

General and travel search engines have both gained ground in 2008. Sixty-seven percent of travelers report usually visiting general search engines in 2008, versus just 53% in 2007 (see Figure 8). Thirty-three percent usually visit a travel search engine.

Almost half (48%) of travelers usually visit an online travel agency first when planning a leisure trip (see Figure 9). Twenty-two percent visit a general search engine first and only 14% visit a hotel, airline or car rental Web site. Travelers over age 55 are significantly more likely than other age groups to visit a supplier site first and less likely to visit an online travel agency.

Travel review sites are generally visited later in the shopping process. Just 4% of online travelers report visiting those sites first, despite the fact that 29% report usually visiting them. Surprisingly, travel search engines, which serve the primary function of comparing prices (the most common reason for visiting multiple sites), are only visited first by 8% of travelers; this suggests that travelers are not comfortable with visiting just one site to compare prices, even if that site is a metasearch site.

Not only do travelers tend to visit online travel agencies first while planning leisure travel, but they also tend to visit them when booking their trips. Fortyseven

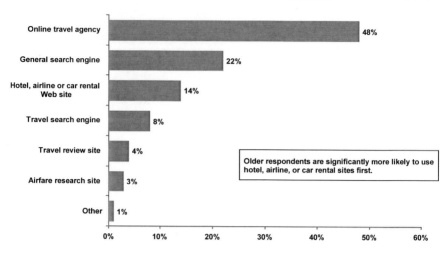

Note: Q4. Which of the following types of Web sites do you usually visit FIRST in the course of planning a leisure trip?

Fig. 9. Web site usually visited first when planning travel

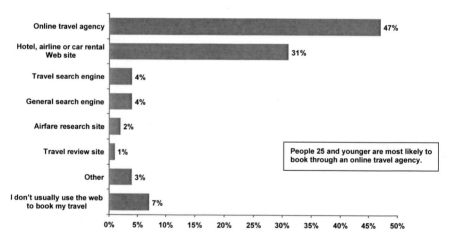

Note: Q6. Which of the following types of Web sites do you usually use to BOOK your travel? *Check all that apply.*

Fig. 10. Web sites usually used to book travel

percent of travelers usually book using an online travel agency, while 31% usually book using a supplier site (see Figure 10). Online travel agencies have been particularly successful at attracting young adults; 61% of travelers ages 22-25 usually book at those sites.

In previous research, travelers exhibited uncertainty regarding where their travel booking was actually taking place, and some travelers consistently stated that they book through general or travel search engines. Many types of travel Web sites now contain a travel search interface despite the fact that they do not actually book travel, which serves to obscure the point of sale. Travelers were intentionally given the option of stating that they generally book on these non-transactional sites, and only 4% report booking via either a general or travel search engine; 1% report booking via a travel review site. This may indicate that online travelers are becoming savvier.

The increasing tendency for online travelers to most often begin their travel planning at an online travel agency site and to book there is notable, particularly given the popularity of online travel agencies among younger travelers. However, travelers continue to seek out multiple Web sites to compare prices and access supplemental information, thus providing a continuing opportunity for search, media and supplier sites to attract and potentially convert wandering eyes. The next chapter will address which online travel technologies are most likely to influence travel purchasing decisions – a factor that should play a key role in online strategy given the increasingly competitive battle for online travelers.

3   The Influence of Consumer Technology on Travel

Travelers are reading user-generated travel reviews online – but seem to be taking traveler advice with a grain of salt. U.S. online travelers rate consumergenerated reviews as less reliable than expert reviews, advice from friends, and recommenda-

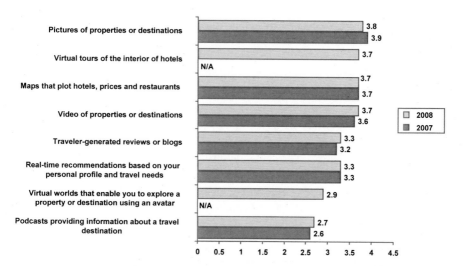

Note: Q12. How influential might each of the following features be when making your travel purchasing decisions? (Means, 1 being "Not at all influential" and 5 being "Most influential.")

Fig. 11. Consumer technologies – relative influence

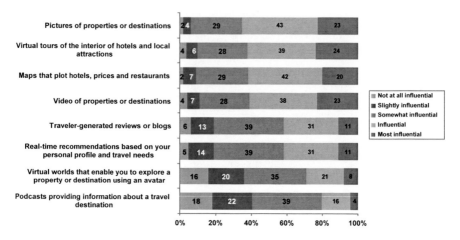

Note: Q12. How influential might each of the following features be when making your travel purchasing decisions? Please rate on a five-point scale, with 1 being "Not at all influential" and 5 being "Most influential."

Fig. 12. Consumer technologies influencing purchasing decisions

tions from traditional travel agents. Eighty-four percent of travelers have read a review of a product or service; just 38% view them as reliable.

Rich media is the most influential type of consumer technology, easily trumping social media like blogs and traveler reviews (see Figure 12). Pictures of properties or destinations are most influential, followed closely by virtual tours, online maps and video. Over 60% of U.S. online travelers find these tools to be influential or most influential when making travel purchase decisions. In contrast, just 42% feel the same way about traveler reviews and blogs.

These traveler attitudes are remarkably consistent with 2007 findings, suggesting a strong and reliable preference among travelers for previewing travel destinations, as opposed to simply hearing the opinions of others (see Figure 11). The least influential technology continues to be podcasts, followed by virtual worlds, which were not assessed in 2007. Notably, virtual tours, which can provide a 360-degree view of a hotel room or resort, are much more influential.

Table 1. Influence of traveler reviews decreases with age

	18–21 Mean	22–25 Mean	26–34 Mean	35–44 Mean	45–54 Mean	55 or Older Mean
Traveler-generated reviews or blogs	3.3	3.5	3.5	3.2	3.2	3.0

Note: Q12. How influential might each of the following features be when making your travel purchasing decisions? (Means, 1 being "Not at all influential" and 5 being "Most influential.")

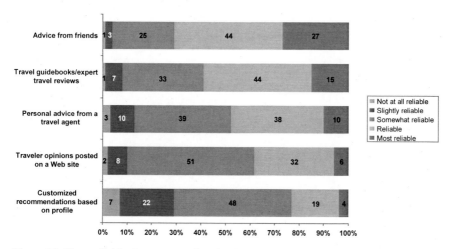

Note: Q9. How reliable do you regard each of the following as a source of travel recommendations? Please rank on a five-point scale, with 1 being "Not at all reliable" and 5 being "Most reliable".

Fig. 13. Evaluating travel recommendations

The general pattern of rich media being most influential holds true for all age groups. However, the influence of traveler-generated reviews and blogs decreases with age, proving most influential to people ages 22–34 (see Table 1). In fact, just 32% of travelers 55 and older rated traveler reviews and blogs as influential or most influential.

Travelers of all ages find advice from a friend to be the most reliable source of travel information (see Figure 13). This tendency has important implications for travel application development within social networks, where highly visible connections to friends and family could stimulate more trusted traveler reviews. Additionally, the high level of trust for recommendations from known associates suggests that social search, which ranks search results based on the preferences of people within one's social network, could play a powerful role in influencing travel purchasing decisions (see the PhoCusWright Technology Edition article "Social Search Emerges in the Travel Vertical" for more information).

Thirty-eight percent of travelers rate travel opinions on a Web site as reliable or most reliable, versus 71% who give the same assessment of advice from a friend. Traveler reviews on the Web are also less reliable than both travel guidebooks/expert travel reviews and personal advice from a travel agent. Notably, a recent study published in PhoCusWright's Global Edition, "Search, Shop, Buy: Inside the Tangled Web of Online Travel," revealed that of the 14 travel reviews sites that ranked among the top 200 most-visited travel Web sites in February 2008, just three were dedicated to consumer reviews. However, while a smaller number of consumer reviews sites have gained traction, by far the most popular reviews site of any kind is TripAdvisor, which was the 13th most popular travel Web site in the same month.

Table 2. Evaluating travel recommendations – influence of age

	18–21 Mean	22–25 Mean	26–34 Mean	35–44 Mean	45–54 Mean	55 or older Mean
Advice from friends	4.0	4.0	4.1	4.0	3.8	3.8
Travel guidebooks/ expert travel reviews	3.8	3.6	3.6	3.7	3.7	3.5
Personal advice from a travel agent	3.6	3.4	3.4	3.3	3.3	3.5
Traveler opinions posted on a Web site	3.3	3.4	3.5	3.4	3.2	3.1
Customized recommendations based on profile	3.0	2.9	3.1	3.0	2.8	2.8

Note: Q9. How reliable do you regard each of the following as a source of travel recommendations? Please rank on a five-point scale, with 1 being "Not at all reliable" and 5 being "Most reliable."

TripAdvisor's popularity, combined with the tendency for travelers to view expert reviews as more reliable, suggests that travel sites should consider incorporating both types of reviews.

Despite the general assumption that social technologies are dominated by young adults, travelers age 18–21 find both travel guidebooks and personal advice from a travel agent to be more reliable than opinions posted on a Web site (see Table 2). If fact, this youngest group of adult travelers finds travel agent and expert advice to be slightly more reliable than do travelers age 55 and older. This may be due to the tendency for travelers 55 and older to find all sources of travel recommendations to be less reliable than their younger counterparts do – perhaps because of skepticism gleaned from experience.

Rich media and technologies that help travelers to share recommendations with people they know remain the most fertile areas for future development. Travelers want to glean as much visual information as possible before making a travel purchasing decision, as illustrated by the high level of influence of rich visual media of all types. And while travellers rate personalized recommendations based on their profiles unenthusiastically, there is great potential to provide travel search results and recommendations based on the preferences and experience of others in one's social network.

4   Social Media Usage, Influence and Marketing Preferences

The nature of social media – especially traveler reviews and social networks – is highly personal. From user-generated content to Facebook applications, travelers are interested only in the information or promotions that are most relevant to them. Reviews from friends and family are significantly more influential than those from strangers. And generic banner ads in social networks like Facebook? Over 40% of travelers view them negatively, opting instead for giveaways or offers personalized to their interests.

Travelers are much more likely to read reviews of travel products and services online than they are to participate in a travel-related social network. In 2008, 84% of travelers reported that they have read a travel review online, while just 20% have participated in a social network – behavior that is nearly identical to the incidence reported in 2007 (see Figure 14). Travelers continue to be much more likely to read reviews than to create them, with 40% reporting that they have posted a travel review.

Hotel reviews are the most commonly read type of travel review online, with 91% of travelers who have read reviews online reporting that they have read a hotel review (see Figure 15). Travel destinations (85%) follow, with restaurants (74%) and activities (60%) attracting a somewhat smaller audience. Of travelers who

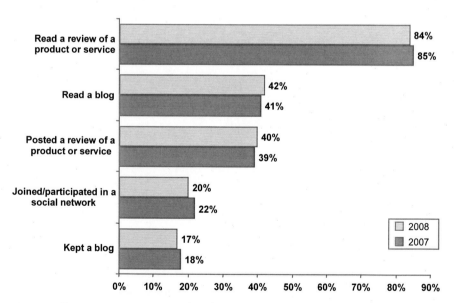

Note: Q8(ii). Which of the following activities, if any, have you done when related to travel? *Check all that apply.*

Fig. 14. Using social travel technology

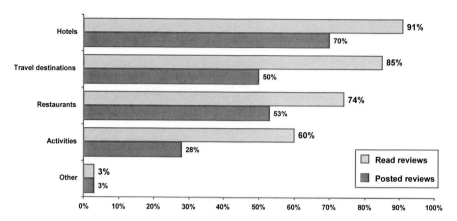

Note: Q8A. If you have read travel-related reviews, what types of travel products and/or services have you read reviews of online? Q8C. What types of travel products and/or services have you posted reviews of online? *Check all that apply.* All data is from 2008.

Fig. 15. Types of traveler reviews

have posted reviews, 70% have done so for hotels, while roughly half have done so for restaurants and travel destinations.

While some industry observers have posited that travelers are most likely to write a review when they have a negative experience to report, in fact the reverse is true (see Figure 16). Just 7% of travelers who post travel-related reviews report writing reviews that are usually negative; in contrast, 50% write reviews that are

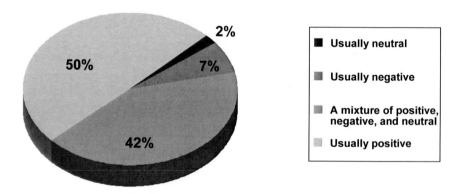

• **Half the respondents (50%) usually post positive reviews**
• **An additional two-fifths (42%) post a mixture of reviews**

Note: Q8D. If you have posted travel-related reviews, were the travel reviews you posted usually positive, negative, neutral, or a mixture of all three?

Fig. 16. Using online technology for travel-related activities

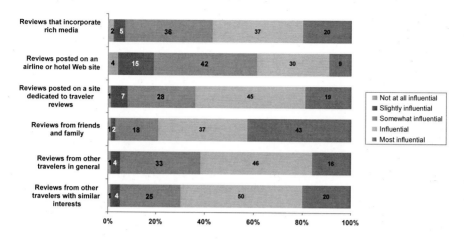

Reviews that incorporate rich media: 2 | 5 | 36 | 37 | 20

Reviews posted on an airline or hotel Web site: 4 | 15 | 42 | 30 | 9

Reviews posted on a site dedicated to traveler reviews: 1 | 7 | 28 | 45 | 19

Reviews from friends and family: 1 | 2 | 18 | 37 | 43

Reviews from other travelers in general: 1 | 4 | 33 | 46 | 16

Reviews from other travelers with similar interests: 1 | 4 | 25 | 50 | 20

- Not at all influential
- Slightly influential
- Somewhat influential
- Influential
- Most influential

- Two fifths of respondents (43%) consider reviews from friends and family "most influential"

Note: Q8B. How influential are the following types of traveler reviews? *Please rate on a five-point scale, with 1 being "Not at all influential" and 5 being "Most influential."*

Fig. 17. Using online technology for travel

usually positive. The remainder tend to write neutral reviews (2%) or a mixture. This trend suggests that while social media makes travel providers' failings highly transparent, it may more frequently serve as a medium for travelers to report positive experiences and promote a favored travel location or brand.

The most influential travel reviews are those from friends and family, with 80% of travelers citing them as influential or most influential (see Figure 17). Outside of their personal social network, travelers are more highly influenced by other travelers with similar interests than they are by reviews from travelers in general. In addition, the type of site on which travel reviews appear makes a difference in how influential they are. For instance, 64% of travelers find reviews on a site dedicated to traveler reviews to be influential or most influential, while just 39% felt the same about reviews posted on an airline or hotel Web site.

This varying perception suggests that traveler review sites like TripAdvisor are perceived as more impartial than sites that are primarily travel vendors (despite the fact that TripAdvisor is owned by the same parent company as Expedia.com). While travel reviews integrated into supplier and online travel agency Web sites add convenience, that value is tempered by an apparent desire among travelers for advice from a third-party source.

While just 20% of travelers report having participated in a travel-related social network, 64% have participated in an online social network of any kind. Of those travelers who have participated in either a travel-related or general social network, MySpace is the most popular, with 64% having joined, followed closely by Facebook at 57% (see Figure 18). At 21%, TripAdvisor (which has added a social networking component) is the most popular travel-related site. Travel-focused social

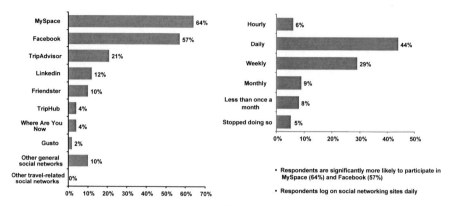

Note: Q8E. If you have joined/participated in a social network – general or travel-related – which of the following social networks, if any, do you participate in? *Check all that apply.* Q8F. How often do you log on to your preferred social network?

Fig. 18. Using online technology

networks TripHub, WAYN (Where Are You Now?) and Gusto combined were visited by 10% of travelers who participated in a social network.

Travelers are most likely to log onto social networks once a day, with 44% reporting that frequency. Nearly 30% log on weekly, while nearly 20% log on monthly or less. Not surprisingly, younger travelers tend to log on more frequently than their older peers. Twelve percent of 18– 21 year olds log on hourly; 63% log on daily. In contrast, for example, 25% of 45–54 year olds log on daily.

Efforts to market within social networks have met with some pushback (e.g., Facebook Beacon), but travelers may welcome certain types of promotions (see Figure 19). Promotions that offer the chance to win free products were rated highest, with 17% viewing giveaways very positively. Personalized offers are the next most appealing type of promotion, followed by viral marketing tools and branded applications. Banner ads are viewed most negatively, with nearly 43% perceiving them as somewhat or very negative.

These trends suggest that relevance, rather than a complete level of privacy, may be the key to advertising within social networks. Travelers are more likely to welcome offers that provide them with a useful benefit, as opposed to traditional banner advertising that more transparently exists solely to promote a brand. The same concept holds true in relation to traveler reviews – a one-size fits all approach to user-generated content will no longer suffice. Social networks and reviews sites that foster travel information-sharing within one's social network and travel sites that successfully sort and analyze user-generated content to provide travelers with only the most personally relevant results are likely to attract the most traveler interest.

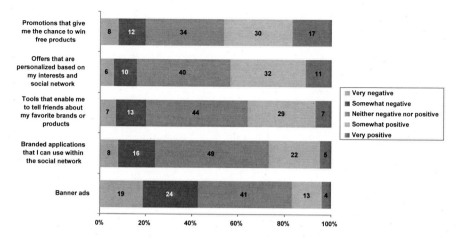

• Promotions for free products and personalized offers are the best ways for commercial enterprises to interact with social networks
– Yet many people (34–41%) are ambivalent

Note: Q8G. How do you feel about the following methods for brands and commercial enterprises to be involved in social networks? *Please rate on a five-point scale, with 1 being "Very negative" and 5 being "Very positive."*

Fig. 19. Travel recommendation

5 Mobile Device Activities and Interest

Innovative mobile handsets like the iPhone – and an increasing number of available mobile Web sites and applications – have led users to upgrade their phones and inspired travel companies to begin devising mobile strategies in earnest. Travelers are gradually increasing their use of mobile phones for activities beyond making phone calls; text messaging, video viewing and Web browsing all gained ground in the past year. In the short term, applications that provide location- or time-sensitive information are the best bet for travel companies introducing mobile services. While travelers express some interest in mobile advertising and booking, these more advanced applications are currently viewed as less desirable than maps, directions and flight status alerts.

Sending and receiving text messages has increased in 2008, with 64% of travelers reporting doing so, versus just 53% in 2007 (see Figure 20). This increase suggests that U.S. travelers may be following the lead of mobile users in Europe and Asia, where text messaging is already widespread. U.S. travelers are gradually increasing their use of mobile phones to access the Internet, with 34% having done so (versus 31% in 2007). Viewing video is also increasing, while m-commerce for both travel and non-travel purchases has not yet gained traction (due in part to the limited availability of m-commerce opportunities). Overall, just 30% are now using their phones solely to make calls, a significant decrease versus 2007 (41%).

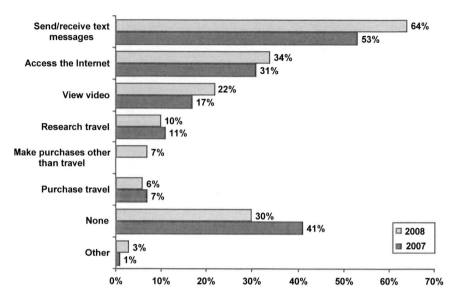

Note: Q13. If you have a mobile communication device, what activities, other than making phone calls, have you done on your mobile device(s)? *Check all that apply.*

Fig. 20. Mobile device activities

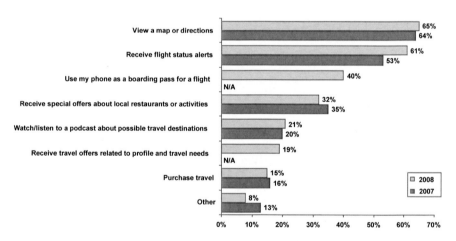

Note: Q14. Which of the following do you think you might be interested in doing on a mobile device? *Check all that apply.*

Fig. 21. Desired mobile device activities

Travelers continue to be most interested in viewing maps or directions (65%) and receiving flight status alerts (61%) (see Figure 21). Notably, 40% of travelers are interested in using their phone as a boarding pass for a flight, a convenience that is currently being tested by U.S. carriers at several airports. Travelers are less keen to begin receiving advertising on their mobile phones, although 32% expressed interest in receiving local offers related to restaurants or activities. In the short term, there are only a small number of travel companies offering bookings via mobile devices; just 15% of travelers expressed interest in purchasing travel via their mobile phone, nearly even with those interested in 2007.

Text messaging is most popular among young travelers – 82% of travelers aged 18–21 have set or received text messages, versus just 31% of travelers over 54 (see Table 3). For most other mobile activities, people aged 26–34 are the most common users: 51% in that age group have accessed the Internet and 32% have viewed video. Travelers 45 and older are most likely to use their mobile phones only for calls.

Interest in more advanced mobile activities generally is higher among younger travelers (see Table 4). However, flight status alerts and viewing maps and directions ranks highly even among travelers 55 and older, with 52% in that age group expressing interest in alerts and 55% in maps or directions. Income level also plays a role, with higher-income travelers expressing greater interest in most types of activities (see Table 5). Forty-eight percent of travelers earning over $200,000 per year expressed interest in receiving special offers for restaurants or activities, versus just 32% of those at the other end of the pay scale. Fifty-two percent of the highest group of earners expressed interest in using their phone as a boarding pass, and nearly 20% are interested in mobile booking. This trend suggests that mobile offerings from luxury travel providers are likely to be embraced by their clientele, a high-income demographic that, in the case of mobile, also may be keen to be early adopters.

Table 3. Mobile device activities: age

	18–21	22–25	26–34	35–44	45–54	55+
Access the Internet	36%	43%	51%	43%	23%	13%
Send/Receive text messages	82%	79%	78%	66%	54%	31%
View video	29%	27%	32%	23%	15%	9%
Make purchases other than travel	7%	7%	12%	11%	6%	4%
Research travel	9%	13%	14%	13%	6%	4%
Purchase travel	3%	8%	9%	6%	5%	1%
Other	0%	2%	1%	5%	3%	5%
None	12%	14%	12%	25%	42%	65%

Note: Q10. If you have a mobile communication device, what activities, other than making phone calls, have you done on your mobilie device? *Check all that apply.*

Table 4. Desired mobile device activities: age

	18–21	22–25	26–34	35–44	45–54	55+
Purchase travel	21%	20%	17%	14%	13%	8%
Watch/listen to a podcast about a possible travel destination	21%	22%	30%	22%	17%	11%
Receive special offers during your trip about local restaurants or activities	34%	34%	37%	36%	29%	22%
Receive travel offers related to profile and travel needs	20%	23%	21%	21%	18%	11%
Use my phone as a boarding pass for a flight	43%	38%	54%	43%	37%	26%
Receive flight status alerts	71%	65%	60%	65%	53%	52%
View a map or directions	76%	71%	56%	68%	68%	55%
Other	0%	2%	5%	5%	11%	21%

Note: Q11. Which of the following do you think you might be interested in doing on a mobile device? *Check all that apply.*

Table 5. Desired mobile device activities: income

	Under $50,000	$50,000–$74,999	$75,000–$99,999	$100,000–$149,999	$150,000–$199,999	$200,000+
Purchase travel	14%	14%	14%	18%	19%	19%
Watch/listen to a podcast about a possible travel destination	20%	19%	20%	25%	19%	23%
Receive special offers during your trip about local restaurants or activities	32%	27%	35%	32%	36%	48%
Receive travel offers related to profile and travel needs	14%	17%	18%	29%	31%	26%
Use my phone as a boarding pass for a flight	40%	39%	34%	47%	33%	52%
Receive flight status alerts	63%	55%	60%	61%	64%	77%
View a map or directions	71%	62%	64%	59%	72%	74%
Other	7%	12%	5%	7%	6%	0%

Note: Q11. Which of the following do you think you might be interested in doing on a mobile device? *Check all that apply.*

It remains to be seen to what extent travelers elect to plan and book travel on their mobile phones. Screen size, handset design, mobile Web site and application quality, and security concerns will all play a role in determining what types of mobile activities travelers seek out, and when. In the short term, mobile services that respond to the particular needs of travelers during their trips are likely to be the most popular. As owned mobile devices become increasingly sophisticated, functionality like global positioning systems (GPS) and m-commerce may become more appealing to travelers, opening the door to more personalized customer relationship management and advertising applications and, in the longer term, new mobile revenue streams.

6 Methodology

The purpose of The PhoCusWright Consumer Technology Survey Second Edition is to gather information about how U.S. online travelers are using and adopting technology when shopping for and booking travel. In addition, PhoCusWright sought to develop a benchmarking study and baseline data against which to make future comparisons.

To qualify for participation in the study, respondents had to indicate if they:

1. Have taken a trip by commercial airline and/or stayed in a hotel in the last 12 months; and

2. Have used a Web site to plan their trips.

The PhoCusWright Consumer Technology Survey Second Edition is based on a 22-question Web-based survey, which was conducted the week of March 31, 2008 and received 811 responses. The respondent pool accurately reflects the national averages for gender, age, marital status, education and household income, and can be projected with confidence to the U.S. adult population of online travelers (as defined above).

The error interval for analysis of groups within the respondent population is +/– 5% at the 95% confidence level.

Destination Management

Developing Southern Mediterranean Tourism: The Interface Between Strategy and Sustainability

Campbell C. Thomson and Harald Jahn[*]

1 Introduction

This paper builds on a work originally carried out using FEMIP Trust Fund[1] re-sources to analyse tourism strategies and policies in the nine FEMIP countries plus Turkey[2], i.e. almost all Southern Mediterranean countries. It reviews the current position of the tourism sector across the region, identifying the strengths and weaknesses of each country, and whether the strategies being followed are com-patible with the concept of Sustainable Tourism.

Tourism has a long history in the FEMIP countries and has become a growth sector in most of them. This is primarily due to the region's natural and cultural resources, a desirable climate, and a location close to key markets. These attrib-utes give the countries of the Southern Mediterranean an outstanding potential for further tourism development; creating income throughout the region and distribut-ing it across socio-economic levels. This, on top of its current economic impor-tance to most of these countries, gives tourism the capacity to influence economic development to a greater extent than any other industry: See Table 1.

The particular potential of tourism development has been understood by the governments of several countries, as well as private sector operators and investors. This has led to the creation of facilities and services to meet a variety of market demands. However, large differences are found between the individual countries. Five countries: Egypt, Jordan, Turkey, Tunisia and Morocco, have a substantial

---

[*] The statements made, and opinions expressed, in this article reflect the views of the authors and do not necessarily reflect the position and policies of the European Investment Bank.

[1] FEMIP is an investment fund managed by the EIB on behalf of the European Union. The Trust Fund is made up of donations provided by a number of EU Member States to be used by the EIB for fundamental research into agreed topics of importance to the FEMIP countries.

[2] FEMIP Countries = Algeria, Egypt, Israel, GWB, Jordan, Lebanon, Morocco, Syria, Tu-nisia. These countries plus Turkey make up the MEDA region.

Table 1. Economic importance of the tourism sector in FEMIP countries

Country	% of Export	Rank	% of GDP	Rank
Algeria	9.3	8	1.8	9
Egypt	31.8	2	8.5	2
Israel	7.2	9	2.4	7
Jordan	22.8	5	6.3	4
Lebanon	52.1 [1]	1	3.6	6
GWB	N/A	N/A	N/A	N/A
Morocco	29.6	3	9.5	1
Syria	24.1	4	2.1	8
Tunisia	22.1	6	5.0	5
Turkey	20.9	7	8.8	3

Source: WTTC
Note: [1] Figure is distorted by the special political situation in Lebanon

tourism development, while four others: GWB, Lebanon, Israel and Algeria, have seen lower, or even negative, growth rates. The question is whether the existing facilities, and future plans, will be sustainable in the long term. Is the need for politically desirable short and medium term financial and economic gains being given greater weight than the long term, continuing benefits that a genuinely sustainable tourism industry would bring?

Across the region the average growth rate has been 12.2%, despite political unrest and terrorism having played a negative role in some countries. Tourist expenditure has been increasing even faster. These growth rates are more than double the world average. Nevertheless, total tourist arrivals to FEMIP countries in 2006 only accounted for 6.8% of the world total, and were lower than the arrivals to Spain in the same year. There is therefore still a significant potential for growth in the sector.

The UNWTO gives a generally accepted definition of the term tourism. A *tourist* is:

> "a temporary visitor staying for at least twenty-four hours and less than one year in the country visited for a purpose classified as either holiday (recreation, leisure, sport and visit to family, friends or relatives), business, official mission, convention, or health reasons".

It is always worth remembering that, unlike business tourists, leisure tourists are under no obligation to visit specific destinations or facilities and tend to be price and fashion conscious. They tend to concentrate their activities in specific (holiday) periods, and are influenced by marketing and publicity. The level of leisure tourism is heavily influenced by living standards, discretionary income levels and holiday entitlements. At the present time, November 2008, with the world economy likely to enter a recessionary phase, the almost forgotten question of income elasticity is likely to play an important role in the demand for tourism products.

Business tourism, on the other hand, is more closely related to general economic development, driven by cross-country exchanges of goods and services. Senior level business travel requires similar, or higher, hotel and travel standards than leisure tourism, but usually focused on city centres. However, an often ignored business sub-sector is the travel and accommodation needs of craftsmen and technicians. These often require lower rated accommodation, but for extended periods. The role of business tourism has not been ignored, but the focus of the study is leisure tourism.

The definition of tourism given above may not be perfect, but it is widely accepted as a working definition and has given "tourism" a common meaning, at least within the industry. The term "Sustainable Tourism", on the other hand, is widely used but there is no common meaning either inside or outside the industry. For some, it is synonymous with ecotourism, for others it is low- or zero-impact tourism. At the EIB, we do not have a specific definition for sustainable tourism. However, we do have a definition for the general principle Sustainable Development. This is in line with the EU definition and is applied to all projects. At its simplest, the definition reads:

> "Sustainable development means that the needs of the present generation should be met without compromising the ability of future generations to meet their own needs"
>
> > (World Commission on Environment and Development, The Brundtland Commission, 1987)

This definition might be interpreted as having only a physical, and possibly social meaning; although we apply it in its widest sense. However, as a lending institution, we also need to be sure that the projects we finance are also sustainable financially and, as the "House Bank" of the European Union, we need to be sure that our projects are economically sustainable.

This Paper will concentrate on the physical, social and financial aspects of sustainability and how it might, or rather how it should, influence tourism investments in the coming years.

2   Characteristics of Tourism in the Region and Beyond

The actual and potential strength of the FEMIP countries is primarily due to their endowment with a wide range of natural and cultural tourism resources: climate, beaches, mountains, coral reefs, deserts, oasis, clean water, food, music, historical and religious sites and attractions, etc. Secondly, it has the advantage of being located close to key markets, making the FEMIP countries easily accessible. This closeness is not just physical. For most Europeans it is cultural as well. From schooldays, the impact of the Mediterranean civilisations and the linked histories of Europe and North Africa create a cultural bond, as do the shared monotheistic religions. These advantages have been understood and used by several Governments, as well as private sector operators and investors. As a consequence, tourism facilities and services in these countries have been provided to meet a variety of market demands, allowing them to take advantage of both their existing resources and the potential market.

Of the Southern Mediterranean countries, five are major international destinations and can be said to be driven by the international market: Egypt, Jordan, Morocco, Tunisia and Turkey. Syria and Lebanon are driven by regional visitors, and Israel by domestic clients. Algeria, Libya, and GWB have tourist sectors which are either "on-hold" or relatively small compared with their neighbours, and which are not yet controlled by free and structured market behaviours.

Even for the "Big Five", tourism has not yet achieved its full potential as a development tool. There are numerous reasons for this but the two most obvious are a) the risk of terrorism, and b) institutional systems and structures.

Most of the Big Five have experienced terrorist attacks targeting tourists. Tourists may have appeared to overcome these incidents relatively quickly, e.g. the bombs in Sharm el Sheikh and Istanbul, but the general image of the countries has suffered. This has hampered not only the growth of tourism in the country concerned, but also in neighbouring countries and the region as a whole.

The second issue is much less straightforward, as it is linked to the structure of the tourism sectors in the countries. To a very large extent, tourism development in Southern Mediterranean countries is driven by a combination of:

- Government authorities and investments in infrastructure and public utilities;

- Major international, regional and local investors and banks;

- International hotel management companies such as Accor, Intercontinental, Marriott, and Hilton, and;

- International tour operators.

The FEMIP Trust Fund *"Analysis of Tourism Strategies and Policies in the FEMIP Countries and Proposals for Sub-regional Tourism Development"*[3] took a dualistic approach to the research. The first analytic approach was carried out in all countries and considered 60 indicators under eleven "hard" tourism headings. Each country was given a development rating for each heading, and the results tabulated to produce Table 2.

---

[3]   Available at www.eib.org/attachments/med/tourism_strategies_policies_2007_en.pdf in English or in French at www.eib.org/attachments/med/tourism_strategies_policies_2007 _fr.pdf. or from the authors.

Table 2. Matrix of key tourism development issues

	Algeria	Egypt	Israel	Jordan	Lebanon	GWB	Morocco	Syria	Tunisia	Turkey
Policy	Y/Z	Y	X	X	X/Y	Y	X	Y	Y/X	Y
Plans	Y/Z	Y	X	X/Y	Y	Y	X	Z	Y	Y/X
Products	Z	Z	X	X/Y	Y/Z	Z	X	Y/Z	Y/Z	Y
Market	Z	Z	X	X/Y	Y	Z	X	Z	Z	Y
Access	Z	X/Y	X/Z 1)	X	Z	Z	Y	Y/Z	X/Y	X/Y
Infra.	Y/Z	Y	X	Y	Z	Z	Y	Z	Y/Z	Y
HRD	Z	Y/Z	X	Y	Z/Y	Z	Y	Y	Z	Y
Regul.	Y	X	X	Y	Y	Y/Z	Y	Y	Y	X/Y
Gov. Support	Y	Y	X	X	X/Y	Y	X	Y	X	X/Y
Collabor.	Z	Z	X	X	Y	Y	X	Z	Z	X/Y
Finance	Y	X	X	Y	Y	Y/Z	X	Z	X	X/Y

Note: [1] While Israel is entirely open to nationals of some 65 countries it regards as friendly, potential tourists from all other countries require a visa, which will be treated on a case-by-case basis. This includes all other FEMIP countries, all countries in the middle-east, and all former states of the USSR except Russia.

[2] There is no rating for Libya. Libya does not fall under an EU-Mediterranean mandate, and is not eligible for FEMIP support.

Legend: X: Systems and actions generally in place
Y: Systems and actions in various stages of implementation.
Z: Actions are required to remove some or several bottlenecks

The second approach was more qualitative: a "soft" approach. Tourism is a global product and the FEMIP countries cannot be seen isolated. The factors affecting global tourism will have a strong impact on the region. One example is the climate change. Emissions from air travel, and the true cost of fuel, will have a long-term effect on travel cost and thereby on travel patterns.

Tourism accounts for approximately 10% of global GDP, with globalisation moving power away from Governments and into the hands of the private sector. While there are benefits from this trend, there may be negative effects, e.g. on the environment. Governments may be in a weak position, with local travel and tourism becoming reliant on international tour operators and developers. A greater partnership between public and private sectors at local and national levels will be required to establish a win-win situation with local and national interests. Tourism sector focused government regulation and intervention will be required, and Governments have to understand the importance of travel and tourism within the eco-

nomic, social and environmental sectors. E-tourism, and the influence of technological changes on the structure of the tourism industry, forms another key issue.
The understanding of the importance of the internet, and dynamic packaging, are
vital issues. While tourism has emerged through mass marketing in the past, developments in technology are creating more custom-made and individual tourism.

Tourism, as a consumer product, has features which separate it from other
products. For example, in other sectors, the delivery of goods, and even services,
is increasingly taking place at the locus of the consumer, even if the consumer is
mobile. In tourism, the delivery of the product almost always takes place at the
point of production: the product is fixed; it is the customer who has to travel. One
exception might be network events, such as golf tournaments which rotate round
different venues. Delivery of the same product then takes place in different locations. However, the tourist still has to be mobile.

Technology is leading to an increasingly consumer-centric approach, with
businesses seeing tourists in a different perspective. Attention is moving away
from destinations as "the product", towards activities such as golf, shopping, spa
and health, food, cultural events, etc. Tourism has therefore become more a matter
of understanding individual consumer behaviour and desires. New avenues of distribution and networking are required and the focus lies in maximising distribution, and accessing the diversity of consumers, to provide a full and detailed
knowledge of the product to the individual.

The concerns over sustainability and the development of partnerships/net-
working are found to be directly related to the development of SMEs. Although
SMEs have limited influence and almost no voice in tourism development, in
reality the tourism sector is highly dependent on the diversification and availability of SME products. This dilemma is not fully appreciated in most countries. However, with the development of technology, SMEs can now reach the
international market place and thus compete with the big operators. Similarly,
many countries underestimate the economic importance, the value-added, and
knowledge created through the backward linkages from the tourism industry,
most of which lie in SME territory.

There is an increasing demand for new products in tourism. Lower cost no
longer means low service and low quality. Attention has moved towards providing
high quality products at a reasonable price. Tourism has become a mature industry
with a focus on what the tourist wants and with an emphasis on innovation as an
instrument for creating new and memorable experiences, on selling stories, on
touching, and on genuine products. The demand for diversity will lead to new opportunities: not just through new products but, more crucially, through new forms
of packaging. This change in direction could be of a great benefit to SMEs. They
can more easily provide memorable experiences, as they themselves are normally
part of the experience. The individual environmental impact of SMEs tends to be
less significant and more easily managed. Developing the role of SMEs in tourism
could therefore not only have significant financial and direct economic benefits, it

could also enhance the sustainability of tourism services and generate additional indirect economic benefits.

While the above issues were developed for the specific needs of the FEMIP study, it is worth noting that there is a substantial overlap with UNWTO target areas of interest.

3   Tourism Trends in the Region

All FEMIP countries have product diversification as a key strategy, with the aim of increasing competitiveness and added value. The objective is to develop higher income market segments and increase tourism employment. New products under development include cruise shipping, thermal and spa (health) tourism, ecotourism and golf activities. However, countries such as Egypt, Turkey, Algeria, and Tunisia still emphasise the "Sun, Sea and Sand" product.

Another important trend is the combination of real estate business and tourism. This is particularly seen in Turkey, Egypt, and Morocco, and is developing in Algeria. However, Tunisia has barriers which effectively prevent this type of development. The market for retirement or second homes is very high in locations with: good climatic conditions, relatively low costs, investor security, and high levels of services, facilities and activities. While holiday home acquisition was previously only for the wealthy, and limited to Spain, Italy and southern France, there has now been a democratisation of demand, with attractive possibilities for FEMIP states. On the other hand, financial resources for investment in beach-oriented tourism resorts are relatively easy to find, while the significant investments needed in wider infrastructure: air, land and sea transport, water, electricity, etc., or the protection of cultural resources, make real estate developments more difficult.

A serious constraint to growth is the status of the promotion, marketing and branding programmes of the individual countries. A good marketing and promotions programme has to be based on: sound statistical information, collaboration between the private and public sectors through efficient institutional arrangements, and sufficient joint financial resources. Weaknesses in these three areas are resulting in the delivery of ineffective marketing and promotional activities. However, effective collaboration between the private and public sectors requires a further liberalisation of the tourism sector in almost all FEMIP countries. The main driving force in tourism marketing is the private operators, and their participation is crucial to the success of tourism development and marketing programs. These need co-ordinated regional tourism-product branding to allow the consumer to differentiate between the leading leisure destinations at the global level.

Secondly, HRD development is a particular problem in the tourism sector. Most of the countries have official systems in place, but these are often focused on management teaching at university level. Hands-on, vocational, and language training, which are essential for the tourism industry, are usually inadequate. This is another case where collaboration between the private and public sectors is vital, and most countries can learn from best practices within the FEMIP countries, or worldwide experience.

Optimal tourism development is based upon an apparently simple and straight-forward process:

- *Government commitment;*

- Overall tourism *development strategy* (e.g., master planning), followed by;

- *Detailed sector planning* (e.g., regional planning, public infrastructure (transport/electricity/water/waste treatment), action programmes, human resources and strategies for investment, and product diversification), combined with an;

- Adequate institutional set-up building upon private and public sector partnership.

However, this roadmap for success is often not fully understood by the FEMIP countries, and various types of divergence can be found in individual countries.

The advantages of the region have already been described. When inexpensive labour, a relatively unspoilt environment, and high standard facilities and services are added, unique tourism products can be created and marketed. Overall, the FEMIP region has great potential to become one of the world market leaders. There is competition from traditional tourist receiving countries in Europe such as Spain, Italy, France and Greece, but many of the European mass tourism products are close to saturation, beaches are overcrowded, resorts are getting old, and costs are high. The historic trend of tourism in Europe has been to look to "the south" for the sun, the sea and the sand. The trend has now gone further south: to Gran Canaria and Senegal, and to the FEMIP countries, although there is also competition from Eastern Europe and long-haul destinations: Asia, Caribbean, etc. This trend is likely to continue and demand for the tourism products and facilities of the FEMIP region is expected to remain high in the coming years. Additionally to its traditional client bases in Western Europe and the Middle East, there is the prospect of tourists from Eastern Europe and Asia. Seen from a competition point of view, the FEMIP countries have a unique market situation, with a product scope which can extend beyond the current products to include, for example, city tourism. There is therefore the potential for significant growth in the number of tourist arrivals, and for optimisation of tourism revenues. To achieve all this will require a high degree of professionalism, sophistication and development, and vision. At present, not all FEMIP counties have these characteristics.

In general, tourists are heavily influenced by security considerations, and family travellers in particular try to avoid putting their families at risk. Safety issues will usually outweigh the loyalty leisure tourists feel to a specific destination. All FEMIP countries have experienced, or been affected by, political violence, terrorism and lack of stability to a greater or lesser degree. Almost all of the countries have suffered one or more negative event directly. However, the most affected areas are GWB, Algeria, Israel, and Lebanon. The tourism development targets outlined below are based on the assumption of a calm and peaceful situation, where tourism can revert to historical trends. Algeria and Morocco both suffered terrorism attacks during the analysis phases of the original FEMIP study, and Turkey, Egypt, and Jordan have experienced terrorist incidents affecting international visitors as well as nationals. While the tourism industry has recovered relatively quickly from such "incidents" in the past, some observers predict that there are politically unstable situations underneath the surface in some of these countries which might change that situation. There is a growing conflict between moderate/progressive forces and religious fundamentalists in all FEMIP countries. Undertones of this conflict can be felt even in moderate and progressive countries like Tunisia, Morocco and Egypt. As long as the unrest in the region persists, and the underlying problems remain unresolved, there will be a continuing possibility of tourism in the region being badly affected.

FEMIP countries have generally had very strong tourism growth over recent years; stronger than in almost any other part of the world, despite the political tensions referred to above. This growth is market driven but based on the tourism resources previously described. The development which has already taken place is just a start and tourism growth can be expected to accelerate in the coming years. Table 3 summarises the growth projections.

There are many similarities in products, and competition between destinations, within the region. For example, inexpensive beach holiday packages offered in Turkey compete with: similar products in Egypt, family beach holidays in Tunisia, some of the new Moroccan beach products, and possibly, in future, beach products

Table 3. Predicted tourist arrivals in FEMIP countries 2006 to 2010

Country	Tourist Arrivals 2006 Millions	Annual Average Growth % 2001-2006	Tourist Arrivals 2010 Millions	Annual Average Growth % 2006-2010
Algeria	1.4	11.2	2.7	14.4
Egypt	9.1	14.8	12.2	7.5
Gaza/West Bank	Na	Na	Na	Na
Israel	1.8	13.3	2.4	10.7
Jordan	3.2	6.5	4.7	10.0
Lebanon	1.1	11.7	1.6	14.9
Morocco	6.6	8.5	10.0	19.4
Syria	8.0	24.0	12.6	16.3
Tunisia	6.6	5.4	8.7	7.0
Turkey	19.8	15.5	33.2	15.0
Total	57.6	12.2	88.1	11.6

in Algeria. In this field, more and more developed, integrated and sophisticated tourism resorts will be offered to the market. Turkey, Egypt and Morocco have many such projects in preparation, airline access will improve and tourism packages will be offered giving very good value for money. In this product domain, quality and price will be important parameters and tourism destinations like Jordan, with a high cost structure, will have to position themselves differently, and avoid competing in the mainstream market place.

Golf courses are being built across the region and Turkey, Egypt and Tunisia seem to compete against Morocco for being "the" golf destination. Similarly, there are extensive plans for spa and health tourism. Like golf, the market for "wellbeing" tourism is increasing strongly, partly driven by an ageing European population with greater disposable incomes. The different cultural background of the FEMIP countries, the product diversification, the traditions, and the languages will lead to market segmentation and to a total increase in the number of travellers to the region. The importance of intra-regional demand is also likely to increase. The FEMIP members will therefore have to work jointly on "competition": co-operating and competing at the same time.

These trends lead to a focus on a few key parameters:

- The service level – the human resources – staff skills will be more and more important. Product diversification will also depend upon the availability of trained staff.

- Innovation in all areas of development: resort layout and the architectural style of buildings, operations and management using international skills with a local focus and, last but not least, innovation in packaging, positioning, branding and marketing.

- Quality: price relationship – good quality does not necessarily need to be expensive, and an open, friendly atmosphere can be achieved at minimal cost.

- Accessibility – Open sky policies and budget airlines are important if the tourist destinations are to grow without being dependent on the tour operators.

- For some FEMIP countries, future market growth will lie with FIT (Foreign Independent Traveller) type clients. This new generation of travellers, often with sound tertiary education and language capabilities, is using the internet as the demand and sales interface, rather than travel agencies and tour operators.

Tourism is about people but it is also about genuine experiences. Travellers are becoming more sophisticated and are looking for "true products". At the same time, travellers are becoming more experienced. They want to escape from their compartmentalised, everyday work and be actively involved in designing their own, unique tourism products. This is creating new possibilities for SME development and specialised tourism.

4   Sustainability Issues

So far, only the demand side has been considered. However, it is the supply side which actually creates the economic benefit and which has to bear the long term environmental impacts. In considering the issue of tourism sector sustainability it is first necessary to identify those actions and activities which will have long term impacts and hence influence sustainability. Sustainability problems always arise if a) the activity involves consumption of renewable resources at a greater rate than the resources can be renewed, and/or b) if non-renewable resources are consumed. A wide range of resources could be considered, but this paper will only consider four from each category.

The following list of non-renewable resources can be considered as a first step. Some of these may at first appear surprising, but they follow naturally from the tourism activities mentioned above and targeted by all Southern Mediterranean countries:

- **Non-renewable energy sources e.g. coal, petroleum derivatives and the impact of Greenhouse Gas Emissions**. It seems an obvious statement, but tourists are only tourists if they travel. To travel requires energy, and currently most of that energy comes from bio-fuels of fossil origin. Extracting bio-fuels formed in prehistoric times is clearly not sustainable. Unless countries are going to rely on tourists who can walk, cycle or sail to their destination, tourism, particularly in the Southern Mediterranean, can only become sustainable if alternative forms of energy can be identified, developed and applied. There may be a medium term strategic advantage for the region as rising oil costs make long-haul destinations too expensive, but in the long term, the current dependence of the tourism sector on fossil oil is a barrier for becoming a sustainable activity. The same issues apply to energy efficiency in buildings: heating and cooling to insulate the tourist from the environment both have a sustainability impact: new solutions need to be identified and developed.

- **Land.** It can be argued that, in the long term, land used for tourism development today will still be available to the next generation for new tourism development. And it may be true for any individual patch of land. However, tourism developments seldom take place in isolation. When an area is zoned for tourism development, the end-of-life implications of the land are rarely, if ever, considered. The thinking would appear to be that one building in a tourist destination can be knocked down and replaced with another. However, that thinking is naïve, because it assumes that tourism products and demand are consistent over time, or at least progressive. Zoning plans for today may not suit what is needed in 50 years time – but the land has been committed and cannot readily be recovered. There is also the impact on a tourist area of one or more patches of disused land undermining the environmental quality of an otherwise attractive area.

- **History.** A region's history cannot be changed by the present and immediate future – although it can always be rewritten. However, history is not just the oral and written word. It is perceived by tourists through its physical manifestations: archaeological sites, man-made landscapes, buildings, etc. These can be copied and reproduced, but the real history lies with the originals and any degradation of those makes history unsustainable.

- **Culture.** The very act of presenting culture sows the seeds of its destruction. It has to be adapted to suit the audience and in so doing induces irreversible cultural change. The "performers", even when they are passive, such as the residents of a native village which is on a tourist trail, will be affected by the activity, and the culture will change.

The renewable resources which can suffer from over-consumption may be less contentious, but the issues are equally serious and may not be limited to the short term.

- **Water and the Cost of Scarcity.** Water is perhaps the most important sustainability issue for the countries of the Southern Mediterranean. With the possible exception of Turkey, most countries are at best semi-arid, and there are already disputes over water rights involving a number of states and other stakeholders. If the predicted effects of climate change do occur, then the problems will be exacerbated.

- **Food.** The introduction of large numbers of tourists will increase the demand for local food production, not only in line with their numbers, but they may seek produce which is less well suited to local production, reducing the net productive capacity of the country. Particularly in semi-arid areas, expanded production levels may not be sustainable in the long term, reducing future potential production.

- **Labour.** As already noted, tourism accounts for substantial levels of employment, particularly if indirect effects are fully considered. The provision of services to tourists usually requires some degree of seasonal, internal migration. Dependency on a single sector and its specific employment patterns may affect not only the long-term labour market, but also the cultural integrity of the regions as well. The effects will be magnified if there is immigration, and emigration, to meet the demand for tourism services.

- **Money.** Money, in the sense of investment capital, is finite, even when cross-border flows are considered. If one sector drains liquidity from the local financial markets, and even more so from the international markets, then other sectors may not be fully financed and the future liquidity will be absorbed by the service and repayment of external debt. Long-term debt to satisfy short and medium term aspirations may limit the ability of future generations to invest in order to meet their own needs and desires: a prerequisite for sustainable development.

5 Common Strategies

Interpreting the FEMIP study, plus additional research on a number of the countries concerned, shows that, with the possible exception of Israel, there is a high degree of commonality in the tourism strategies in the Southern Mediterranean. There is a well known saying in English "Great minds think alike: fools seldom differ". Who falls into which category? If the Ministries of Tourism are to be given the benefit of the doubt, it may be that they have identified a common set of solutions to a common set of market conditions.

- **Growth Rates.** At the time the FEMIP report was prepared, world tourism was expected to continue to grow at approximately 7.5% per annum. Almost all of the countries concerned were basing their strategic planning on growth rates which were significantly higher, in some cases double the world rate.

- **Target Clientele.** All countries were planning to attract higher value tourists and wanted to move away from mass, low value tourism: but see previous point.

- **Tourist Draws.** All countries were planning to develop tourism which would take advantage of their history and culture.

- **High Value Added Services.** All countries wanted to offer additional experiences over and above Sun, Sea, Sand, History and Culture. Typically these take the form of golf courses, ecotourism and spas.

- **Client Origins.** Almost all countries were planning to maintain their traditional client base. However, they were also planning to target newly developed countries and large countries with high rates of economic growth in Asia. Here there was some variation in specific targets, but still a high degree of overlap.

6 Strategy: Sustainability Mapping

The strategies and key sustainability issues can now be mapped to identify the key areas of potential conflict between the two (cf. Table 4).

At this stage, the ratings above are largely qualitative and experiential, rather than quantitative. It should also be noted that there are a large number of crosslinks, e.g. between the target clientele and additional client services – although few in which a positive impact counterbalances a negative impact. It should also be remembered that potential impacts are presented, and not actual impacts. Finally, there a number of strategies which could have a positive impact on sustainability.

Table 4. Potential impact on sustainability

	Growth Rates	Target Clientele	Tourist Draws	High Value Added Services	Client Origins
N.R. Energy	- - -	+	- -	- -	- -
Land	- - -	+	+	- -	0
History	- -	- -	- -	0	0
Culture	- -	- -	- -	-	-
Water	- - -	0	0	- - -	0
Food	- -	0	0	-	0
Labour	- -	-	-	+	0
Money	- -	+	+ +	- -	+

+++ : Greatest potential positive impact on sustainability

- - - : Greatest potential negative impact on sustainability

  0 : Neutral impact

Rather than considering the forty positions individually, the following analysis is based on the strategy dimension, which should be shorter and more meaningful in terms of the implications for strategic development in the various countries.

6.1   Client Origins

The analysis ignores the potentially different behaviour of clients from different countries and regions of the world. It also ignores the impact of different tourism patterns, e.g. individual, family groups, small groups, large groups, from the various countries and regions. This column has the greatest number of neutral observations, reflecting the idea that a tourist is a tourist irrespective of origin. However, there are three areas where the finding is not neutral. Firstly, there is the energy factor. Historically, Southern Mediterranean tourism has been dominated by short- and medium-haul air travel, plus road travel at the Eastern end. The newly targeted clients either originate from Central and Eastern Europe, with fewer short-haul flights on average, or are long-haul visitors from further afield. In both cases there will be a greater environmental impact. This will not directly affect the sustainability of tourism within any individual country, but it may have an impact on tourism overall, hence the negative rating. Unless, of course, practical sustainable energy sources become available.

Secondly, it can be argued that the damage to the cultural integrity of the destination countries from the current mix of tourism has already happened and the impacts assimilated. Now a new wave of tourists will arrive, some of whom will be even further removed from the local culture. Meeting the needs of those tourists, and their interaction with local communities, may dilute local culture still further.

Finally, a positive aspect is that most of the visitors from the target countries will have lower expectations in terms of accommodation and services. A willingness to accept lower standards would suggest a lower capital investment per visitor, and hence a less negative impact on sustainability.

The paradox of having simultaneous strategies of attracting visitors from relatively less affluent countries and targeting the high-value tourists is ignored for the purposes of this paper.

6.2   High Value Added Services

It should first be stated that not all high value added services have the negative impacts shown. By definition, properly controlled ecotourism would be neutral. The main issues relate to activities such as golf courses and spas.

As the table shows, these activities present significant sustainability issues. Spas and golf courses are energy intensive: in the case of golf courses the energy content of fertilisers, maintenance and irrigation need to be considered. They both require substantial volumes of water. Most spa water can be recycled, but a golf course in a semi-arid region can consume the same volume of water as a town of 30 000 inhabitants. Golf courses change the nature of the landscape and the vegetation, often taking fertile land out of production. However, all of these value added services generate economic activity by attracting tourists who would not come otherwise. They are mechanisms by which the season can be extended: producing a positive impact on labour.

6.3   Tourist Draws

History and culture are intangible, but while history can be rewritten it cannot be changed. As already suggested, culture is fragile. However, both history and culture, and the artefacts which represent them have the great advantage for the tourism sector of being free. They may require some investment to restore and maintain, but they are the classic example of a sunk cost asset which can still provide economic benefit. Hence the positive rating for money and land: assuming that historic sites are protected.

The problems only arise when tourists arrive and start to erode both the culture and the historic sites. Tourists visiting historic sites cause almost inevitable wear and tear, even when it is unintentional. This is what justifies the negative ratings. However, with proper management of the carrying capacity, it may be possible to reach an equilibrium condition: if the resources applied to protection and renovation are adequate, and if the sites are properly managed, it might be possible to achieve a rate of attrition which is no higher, and may even be lower, than the rate which would occur naturally. This applies both to the tangible and intangible.

6.4   Target Clientele

The reference here is the stated target clientele, rather than the clientele which will be required to meet the target growth rates. On balance, meeting the needs of such clients would tend to enhance sustainability, mainly because fewer high value tourists are needed to achieve a given level of economic benefit, resulting in a lower consumption of resources.

The main negative impact comes from their patterns of behaviours, which have an impact on history and culture, and their requirement for higher standards of accommodation, which requires greater investment.

6.5   Growth Rates

The common policy of "going for growth" may create a range of unrecognised sustainability issues. There may be a perception that increasing numbers of tourists will maximise future economic benefits, with a particular impact on job opportunities and reduced unemployment. This is a particular issue in the region, with rapidly growing populations and high levels of youth unemployment. There may also be a perception that this can be achieved at minimal cost to the government. Both of these may be true. However, such an approach, with a high discount rate, might have a long term economic cost which future generations will have to bear.

Most Southern Mediterranean countries have been zoning substantial lengths of coastline and inland water-rich areas as tourist areas, including countries which are not members of the "big five". Setting aside the question of whether these areas are actually needed to meet the probable, rather than possible, demand, questions are being asked about the environmental impact of these developments. In most of the cases in which the European Investment Bank has been solicited for financial support, environmental issues have been properly addressed. It is also true that all of the countries concerned have established a legal framework and enforcement process to ensure that environmentally sensitive areas are protected. However, it should be noted that a number of major developments in the region which have been proposed by private investors, do not, at first sight, appear to comply with national legislation, and may have been accepted by the local authorities on the basis of their economic importance. This issue is widely discussed by some well-respected NGOs.

Clearly, if anticipated growth rates are achieved, there will be major economic benefits to the countries concerned. These benefits could then be used to mitigate the impact of tourism on seven of the sustainability issues listed above. However a risk remains, reflected in the "Money" rating that the physical growth targets will be achieved, but the target growth rate in tourist numbers will not. In that case, resources will have been allocated without the prospect of an adequate financial return, which might create a downward sustainability spiral.

7   Conclusions

The findings of the FEMIP study provide the first structured, trans-regional analysis of tourism strategies in the Southern Mediterranean. As such, it could be a guide and framework for all stakeholders, enabling a dynamic and fruitful dialogue to encourage the development of a tourism industry which combines economic development with long term sustainability. A first step in this process was the first Euromed conference of Ministers of Tourism in April 2008, hosted by the Moroccan government and supported by the European Commission and the Presidency of the European Union. This put the industry, its potential, and its development needs into the wider policy context and was appreciated to the extent that it is to become a bi-annual event.

It will have been noted that the current analysis did not study the price-elasticities of different tourist products, nor included a cost-benefit analysis of the various marketing activities. Future studies might also be directed towards providing a better understanding of the income-price elasticity for both leisure and business related travel and accommodation.

In conclusion, the Southern Mediterranean region is an ideal tourist destination for many European and Middle-Eastern residents. It is relatively close, it is affordable, and it has the natural and cultural assets they are seeking. For the countries of the region, tourism is a source of revenues based on their natural advantages and resources. However, as regions in the Northern Mediterranean know to their cost, intensive tourism development limits the choice of future generations: the basis of the sustainability concept. The current and planned developments are far removed from the Northern problem areas. However, there are two risks for Southern Mediterranean countries. The first is that their strategies will be successful and that the growth in hotel beds will allow the target tourist numbers to be achieved. If that growth were to continue then, ultimately, the industry would cease to be sustainable and there would be a risk of declining revenues and redundant assets.

The second is that the investments will be made, but that the tourist numbers will not be achieved. The investments made will be under-utilised, or costs will be under-recovered, and the industry will again be unsustainable.

The challenge for the Ministries of Tourism is to find the optimal path between these two, and develop a tourism industry which is both financially and environmentally sustainable in a global and regional market place.

Film Tourism – Locations Are the New Stars

Stefan Zimmermann and Tony Reeves

There once was a time when the actual attraction of a movie lied in the capability of the celluloid to let the spectators escape from the everyday, a spare time to leave the life-world environment behind and live the life of a thrilling armchair traveller. Foreign countries and remote regions could easily be mapped and visited without travelling to the actual location. Considering that "the modern world is very much a 'seen' phenomenon" (Jenks 1995:2), one has to come to the assumption that present day travelling and watching movies are somehow connected. Tourism originally was perceived as visitors travelling, whether within their own country or internationally, for pleasure and relaxation. However, over the years, tourism has evolved into different components and labels.

Movies are an integral part of popular culture and everyday routines and therefore impact on many people. Today watching television is the most common home-based leisure activity (Busby and Klug 2001). One might assume that more and more people try to find distraction not solemnly within the imaginary realms of the movies – or in front of the TV set – but nowadays at the actual filming locations. These locations function as a kind of "stargate" where the traveller can enter the realm of his imaginations (Zimmermann 2003). The film location is therefore a perfect place to go to, a place where the cinematic narrative enters the life-world and somehow materializes.

Travellers and travel companies have discovered the power of the visual when it comes to market new needs and channel the tourist perception (Urry [2]2002). We do know that tourists tend to travel to destinations contrasting their everyday environments and it is known that contemporary tourists' images of places are shaped and held alive through consumption of film and TV productions without the perceived tendency of promotional material (Schofield 1996). These environments and the additional imaginary often derive from TV and the cinema in a growing number of cases. The latter is not coming as a surprise, for the relationship of fiction and tourism is very well known and established (Ryan 1997). The media have become a major vehicle of awareness and style leadership in terms of communicating remote environments and spectacular sights (cf. Coates 1991). Therefore it is obvious that film locations tend to hold an idle potential to market a place that is involved in shooting a movie or a TV show. Feature films can enhance awareness of places, regions or even countries and affect decision-making processes. Due to this fact tourism marketers are increasingly working with film producers to promote their destinations as film locations (Seaton and Hay 1998). As film and television consumption continues to expand, one might assume that the overall influence of visual media on place images is growing as well (cf. Kim and Richardson 2003).

This comes along with the fact that tourism – like cinema or watching TV gains part of its attraction from the beauty of the gaze. The spectacular sights turn into fascinating sites and vice versa again. Cinematic narration and touristic staging seem to merge at the most obvious level. Cinema becomes real to such an extent that spectators gain the opportunity to enter parts of the precious imaginary. In this way, film viewing itself may be understood as a form of tourism – a kind of immobile flânerie, which both reflects and constitutes a range of tourist practices. "Places are chosen to be gazed upon because there is anticipation, especially through daydreaming and fantasy, of intense pleasures, either on a different scale or involving different senses from those customarily encountered. Such anticipation is constructed and sustained through a variety of non-tourist practises, such as film, TV, literature, magazines, records and videos, which construct and reinforce that gaze" (Urry [2]2002:3). This gaze is constructed and maintained through signs and it is organized around these signs and symbols, indicating a specific contextual belonging. As Culler (1981:127) puts it: "All over the world the unsung armies of semioticians, the tourists, are fanning out in search of the signs of Frenchness, typical Italian behaviour, exemplary oriental scenes, typical American thruways, traditional English pubs." Tourists and cinemagoers are pretty much the same in that sense that they are looking for established features and already consumed bits and pieces.

Conventional wisdom asserts that, to be popular as a visitor attraction, a location needs to invoke the 'feel-good' factor associated with romance or escapism, yet this does not necessarily seem to be the case. Burkittsville, the setting for *The Blair Witch Project*, has seen an influx of visitors, and the Georgetown house in Washington DC which was the setting for *The Exorcist*, continues to be a major tourist attraction (Reeves 2001).

Perhaps the most surprising example of film-generated tourism is John Boorman's 1974 film *Deliverance*, in which a quartet of city dwellers suffers a series of horrific ordeals in the backwaters of rural Georgia. Despite the grisly fate of the characters, the film sparked a boom in white-water rafting vacations on the Chattooga River in Rabun County, where the film was shot. A much more important factor than screen glamour seems to be a tangible sense of place. Take *The Godfather* for instance – one of the most popular films of all time, yet there's little in the way of tourism associated with it. With its multiple settings, it's hard to think of one striking location image; there is the Godfather mansion, which can be found on Staten Island, but it is not given any visual prominence in the film. Mention of *The Exorcist*, on the other hand, immediately conjures up the poster image of the Father Merrin approaching the forbidding house, with the shaft of light streaming from the upstairs window. This demonstrates that the power of constructed reality is likely to dominate any sense of objective reality (Morgan and Pritchard 1998) and that the visual is only a vehicle to transport the narrative of a movie. Think of *Amelie* and, chances are, you'll remember Montmartre's art deco Café le Deux Moulins. It doesn't occupy much screen time in the film, but the stylised photography and heightened colour carry a visual impact that lingers in the mind. The sense of place, which is constructed by the film narrative, materializes in the visual image of the Café le Deux Moulins and acts as a touristic stargate. Looking at this one might argue that the specific content of the movie can affect the viewer's image and perception of a place portrayed in a film, using both: the visual and the narrative.

Film tourism or film-induced tourism marks a specific development within the field of modern tourism. Media has taken an important role in promoting holiday destinations these days. One might argue that today's travellers have already seen the whole globe and that the urge to discover a thrilling and somehow new spot is one of tourist's major aims, but looking closer on tourism development it is far more than that. Looking at tourism from a cinematic perspective there is no doubt that for a destination "there s no finer publicity than that generated by a major motion picture." (Riley and Van Doren 1991: 267). The term film tourism usually characterizes the effects of cinema and TV productions on travel habits and travel decisions and is therefore a truly measurable media impact (Zimmermann 2003). It also describes every touristic activity concerned with visual mass media. Riley et al. (1998) put it even simpler, they asset that people turn into movie-induced tourists as soon as they are seeking for sights and sites on the silver-screen. Current ways of consuming pictures and images are very close to the consumption of places and landscapes. It appears that movie induced tourism helps to create a new kind of cultural landscape, a conception of landscape that goes beyond known ideas and concepts of history, culture and society. It could further be perceived that movie tourism seems to be strongly connected to nostalgia and identity.

We know that feature films considerably shape people's behaviour and their everyday perception of landscape (Escher and Zimmermann 2001:227), therefore we can be sure that movies also have a profound influence on the perception of places as well. When a movie or popular television series is filmed on-location in any real-life town or region, a growth of visitor numbers can be observed regularly (Riley 1994, Riley et al. 1998, Tooke and Baker 1996, Beeton 2001, Busby and Klug 2001). Usually this is only a first step for further increase of touristic infrastructure (Zimmermann 2003, Escher et al. 2008). We must not forget that movies are usually not produced to visit locations, but the side effect is very well known and sometimes used for marketing purposes (cf. Riley et al. 1998). To be precise the last few years have shown an evident increase in the marketing of film locations and some examples, as most of Visit Britain's projects prove that marketing efforts can be done with the intent of profiting from movie induced tourism. The lesson taught from this is very easy to understand: relationships between film commissions, tourist authorities, film productions and distribution companies should be established. Some places and destinations create organized trips for tourists to see the locations and take the form of paid tours, self-guided tours or by means of location maps as done by Visit Britain since 1996. More than 200 TV and movie locations throughout the UK have been highlighted in this vein (Hudson and Ritchie 2006).

The biggest problem analyzing film tourism seems to be that there is not much profound research on economic gain and visitor numbers so far and neither on reliable statistics. Many locations never counted visitors because they had never been attractions before they gained new prominence through a movie (Riley et al. 1998). Nevertheless the carried out research demonstrated that the use of film locations as travel destinations has considerable value and it is profitable taking steps to lure television and movie companies to a location on the first place and make the best subsequent use of that production (Tooke and Baker 1996).

The extent of movie tourism, though, can be gauged when a location is specifically constructed for a film, within three years of the release of *Field of Dreams* in 1989, it is estimated that 60,000 people had made the trek to Dyersville, Iowa, to visit the baseball diamond built on farmland for the production (Riley et al. 1998:927). In fact, the Lansing farm became the state's top tourist attraction, and is still doing business today.

Anecdotal evidence, too, abounds. After the release of *Gorillas In The Mist* in 1988 (and before the subsequent troubles), tourism to Rwanda increased by twenty per cent. When Steven Spielberg featured Devil's Tower National Monument in Wyoming as the alien landing site for *Close Encounters of the Third Kind*, the number of visitors to the national park is claimed to have increased by three quarters. And, according to Ireland's Minister for Arts, Culture and Heritage, more than one in six visitors to the Republic of Ireland in 1993 cited a reference to the film as their reason for visiting (Reeves 2001).

A widely-quoted example of success following a screen appearance is the Crown Hotel, Amersham, which featured in the hugely successful 1994 romantic comedy *Four Weddings and a Funeral* (Riley et al. 1998). What makes this instance remarkable is that the Crown is never mentioned by name (in the film it's called 'The Jolly Boatman'), and another, more photogenic, inn was used for exterior shots. Nevertheless, fans of the film sought out the location and the suite used for filming became fully booked for several years.

Film induced tourism is of outstanding interest in terms of economic development, as many of the more popular television and film sites can be found in small and often rural communities, sometimes remote areas and therefore might be a possibility to create a unique selling point where without a film production nothing ever would happen. A lot of films are using tourist practises; the way in which places are communicated follows the modes of everyday perception. The tourist gaze is very important to transfer a narrative space into a tourist attraction. It shapes the way in which a cinematic icon is transferred into a life-world place of interest. This process can sometimes be supported by the additional selling of merchandise and memorabilia, functioning as signifier, helping to embed the location within the cinematic world, the tourist seeks to be part of. But still many film locations have not recognized the power of the movies and have missed to use that potential to promote them. The planning of such promotional activities might cause severe problems and have to be planned carefully as Riley et al. (1998:931) argue. Especially when the location differs from the cinematic pattern and the tourist gaze is disturbed heavily. This might be caused by quickly built infrastructure or the sale of cheap memorabilia (cf. Riley et al. 1998). Another problem can be observed for example in Tunisia, where the missing careful treatment of the former shooting sites of the Star Wars movies can be a letdown for many tourists. Or a missing of adequate infrastructure in terms of required accessibility and merchandising opportunities. For good or bad, the interest generated by a film can profoundly change the location itself. Largely unknown at the time of filming, the striking limestone pillar of Khow-Ping-Kan in Phang Nga Bay, Thailand, used as the villain's lair in *The Man With The Golden Gun*, is now not only a major tourist draw but is generally referred to as 'James Bond Island'.

Travellers might visit places that used to be movie locations in the past and still offer an intriguing connection to a specific film, a genre or a TV show. This phenomenon often goes hand in hand with a good portion of nostalgia and can endure over a long period. These locations were often never brought to market as filming locations and the cult about it usually derives from film buffs, turning these places into an almost sacred place of pilgrimage. Some places offer the opportunity to attend the actual shooting and let the tourists be somehow part of the creation of a movie, especially small towns hosting a successful TV production seem to handle this with ease and relaxed attitude.

While the stardom of Hollywood's early years focused on the actors, one can easily see an evolving movement towards stardom of specific cinematic places and landscapes. If the actual location holds a highly recognizable and displayable feature it often turns into a unique selling point and becomes a tourist attraction. This usually happens when the cinematic narration and the specific quality of the depicted spot complement one another. Planning such a success is still one of the trickiest bits, just as the planning of a block-buster movie. There is always the danger of failing at the box-office and there are no guarantees that a film or a TV-production will have a strong influence on the spectators that they wish to visit the place of cinematic origin. But different studies show that the impact of some films to their shooting locations can be tremendous but on the same time proof a measurable influence on these places on an economical, social or somehow cultural level. In general, inhabitants of locations tend to see the benefits a film production might bring to the place, as are short-term increase of employment and a gain of publicity (cf. Busby and Klug 2001).

In some ways, it's understandable that the phenomenon of movie tourism has taken so long to develop. During the early years of the industry, technical limitations meant that films were photographed largely inside a studio. The viewer's experience remained essentially theatrical as actors performed, as they always had, in front of purpose-built – and obviously artificial – sets. To establish the locale, a standard Hollywood practice was to employ a 'Second Unit'. For a film such as *All About Eve* (1951), which was set in the theatre district of Manhattan, the main cast filmed their scenes in the easily controlled environment of a Hollywood soundstage, while a skeleton crew was despatched to New York to capture brief 'newsreel-style' shots of East Coast exteriors.

There was, subsequently, little sense of a 'real' place to trigger the desire in the viewer to visit the places where the action was supposed to take place.

From the late 1940s, developments in lighting, sound and camera technology gave a flexibility that catered to a demand for more realism in films. In 1949, Carol Reed chose post-war Vienna as the backdrop to *The Third Man*, with striking results. Images of the city's squares, the Central Cemetery and – most of all – the ferris wheel linger in the mind. The success of the film established an image of Vienna, which we still hold and, as fans of the film began to follow the footsteps of Harry Lime around the city, it proved to be one of the earliest examples of cinema-generated tourism.

When John Ford made the Oscar-winning *How Green Was My Valley* in 1941, the South Wales locale was recreated entirely in Los Angeles, but eleven years later, he chose to film his whimsical Irish comedy drama *The Quiet Man* on location. Photographed in lush colour, the unmistakable sense of a real place again provided an incentive for audiences to visit the quaint, romanticised 'Emerald isle' they'd seen on screen. The village of Cong, in Connemara, where most of the film was shot, experienced an influx of visitors, which continues to this day. Well over half a century later, Cong's Quiet Man Hostel still shows the film nightly, and offers guided tours of the locations.

Artificial by their very nature, musicals remained one of the most set-bound of genres, but in 1965, Robert Wise broke free of the confines of the studio to film much of *The Sound of Music* in Salzburg. The results are legendary. Coach tours continually ferry tourists from one familiar location to the next, and at least one enterprising tour company has offered charter flights over the castles and lakes of the opening aerial shots.

At the time, though, these films were exceptional. They each commanded legions of loyal devotees in an age when going to the cinema was still an 'event' and only true fans saw a film more than once.

The Seventies saw two developments that rendered the advent of movie tourism inevitable: apart from cheaper and more accessible air travel, the decade heralded the birth of home entertainment.

Amassing a video/DVD collection has fundamentally changed our relationship to cinema. Before the Seventies, a film would rarely be viewed more than twice except by the most diehard of fans – once on its initial release in the cinema and again when it premiered on television. Now a collection of favourite films is the norm for most people, and movies are played in the way once reserved only for music, with films neatly divided into chapters, like album tracks, to which we can skip, and replay, favourite scenes. We are far more familiar with, intimately connected to, our favourite films than at any time in the past.

Busby and Klug (2001:329) explain tourism in an intriguing and holistic matter: travel is therefore a complex, symbolic form of behaviour through which the tourist is usually striving to fulfil multiple needs, leading to the fact that tourists are often unaware of the real reasons and motivations for their travel behaviour. Studies, as carried out by Escher et al. (2008) show that this behaviour often is the result of cinematic adaption or even in a cinematic historiography, with the result that cinematic origins often cannot be seen by a not informed or only poorly informed recipient.

From the point of view of film tourism, a more problematic film is *The Bourne Supremacy*. A huge international hit, the plot seemed to wander all around Europe, but most of the film was in fact shot in Berlin. There was little publicity surrounding this, which raises interesting questions: would it actually have damaged the feel of the film, knowing that the locations were not as far-ranging as they appeared? And would fans of the film be as interested in visiting locations passed off as 'Munich', 'Moscow' or 'Naples' as fans of *The Third Man* are in visiting the 'real' Vienna?

What does that say for cities such as Toronto and Vancouver – and lately Prague and Budapest – which are major filming centres but rarely appear as themselves on-screen? Is it possible – or even desirable – to raise their profile? Or would a high profile merely encourage film-makers to move on to other, less recognisable, locations?

This also raises the question: just what is it that tourists visit? When vacationers flocked to New Zealand after seeing the *Lord of the Rings* trilogy, were they travelling to Middle Earth or New Zealand? Were they Tolkien fans, wanting to follow in the footsteps of Bilbo and Gandalf, or winter sports enthusiasts heading for the Whakapapa Ski Field which they read had been used as a filming location? And did it matter? The *Rings* phenomenon seemed to blur the boundary between the real and the fictitious. Entering the town of Matamata, which provided the rolling green hills of the Shires, visitors are greeted with a sign reading 'Welcome to Hobbiton'. This somehow confronts and complicates the oppositional dichotomy of authenticity and artifice and its narrative of conquest acceptance.

The present publications demonstrate that a movie location generally attracts visitors (Riley et al. 1998, Busby and Klug 2001, Zimmermann 2003, Beeton 2005). Corresponding to Kim and Richardson (2003: 232) there is some statistical evidence that a popular motion picture significantly influences destination images and that an increase in popularity of cinematically depicted locations can be observed. Furthermore movies can be an effective tool to change place images and affect the audiences' interest in visiting a specific destination (ibid.). According to Reeves (2001) there is no doubt, that if you film a specific place, the tourists will come to see the location and must be seen as a useful marketing tool for a single destination, a community, a region or even a country.

References

Beeton, S. (2005): Film-induced Tourism. Clevedon, Buffalo & Toronto.

Busby, G. and J. Klug (2001): Movie-induced tourism: The challenge of measurement and other issues. In: Journal of Vacation Marketing. Vol. 7 (4): 316-332.

Butler, R. W. (1990): The influence of the media in shaping international tourist patterns. In: Tourism Recreation Research 15, (2): 46-53.

Coates, J.F. (1991): Tourism and Environment: realities of the 1990's. In: World Travel and Tourism Review: Indicators, Trends and Forecasts. Vol. 1. 66-71.

Culler, J. (1981): *The Pursuit of Signs: Semiotics, Literature, Deconstruction.* London.

Escher, A., E. Riempp, and M. Wüst (2008): In the Footsteps of Jedi Knights and Sea Pirates. Hollywood Movies and Tourism in Tunisia. Geographische Rundschau. International Edition. 4, No. 3. 46-52.

Escher, A. and S. Zimmermann (2001): Geography meets Hollywood – Die Rolle der Landschaft im Spielfilm. In: Geographische Zeitschrift 89 (4): 227-236.

Hudson, S. and J.R.B. Ritchie (2006): Film tourism and destination marketing: The case of Captain Corelli's Mandolin. In: Journal of Vacation Marketing, Vol. 12(3). 256-268.

Jenks, C. (1995): The centrality of the eye in Western culture. In: (ibid.) (Ed.): Visual Culture. London. 1-12.

Kim, H. and S. Richardson (2003): Motion Picture Impacts on Destination Images. In: Annals of Tourism Research, Vol. 30 (1), 216-237.

Meethan, K. (2001): Tourism in Global Society – Place, Culture, Consumption. – Houndmills, Basingstoke.

Morgan, N. and A. Pritchard (1998): Tourism Promotion and Power: Creating Image, Creating Identities. Chichester.

Reeves, T. (2001): The Worldwide Guide to Movie Locations. London.

Riley, R., D. Baker and C. Van Doren (1998): Movie induced Tourism. In: Annals of Tourism Research. Vol. 25 (4): 919-935.

Ryan, C. (1997): The Tourist Experience. London.

Schofield, P. (1996): Cinematographic Images of a City. In: Tourism Management 17 (3): 333-340.

Seaton, A. and B. Hay (1998): The Marketing of Scotland as a Toursit destination, 1985-1996. In: MacLellan, R. and R. Smith (eds.): Tourism in Scotland. London. 209-240.

Tooke, N. and M. Baker (1996): Seeing is believing: the effect of film on visitor numbers to screened locations. In: Tourism Management 17 (2): 87-94.

Urry, J. (²2002): The Tourist Gaze. London.

Zimmermann, S. (2003): "Reisen in den Film" – Filmtourismus in Nordafrika. In: Egner, H (eds.): Tourismus – Lösung oder Fluch?: Die Frage nach der nachhaltigen Entwicklung peripherer Regionen. Mainz, 75-83. (= Mainzer Kontaktstudium Geographie 9).

Marketing and Sales Management

Brands as Destinations – The New Tourism Objective for Chinese Tourists

Monika Echtermeyer

1 European Destinations Competing with Symbols and Brands

Tourist destinations sell the most important product of the world: "Feelings of happiness." Chinese tourists in particular are desperate for experience when they take their first trip to Europe; a trip which they consider to be highly prestigious. In Europe they look for feelings of happiness and a romantic atmosphere. First-time visitors try to visit as many destinations as possible in just a few days, wanting the "maximum kick" in the minimum amount of time.

The traditional trip across Europe, visiting ten countries in up to 15 days, will continue to be offered in the foreseeable future (given the ever-growing market for first-time Europe travellers). This leaves little time for individual destinations and very little time to linger. Nowadays however, all big tour operators offer single-destination trips and strive to expand the FIT (Free Independent Traveller) sector. In the three largest source markets, Beijing, Shanghai and Guangdong (see Fig. 1), trips taking in just three to five countries have already become the most dominant. In addition, the MICE sector has been growing as globalisation has led to the use of modern incentive methods in China.[1] All of the tourist destinations preferred by the Chinese have one thing in common; sights which the Chinese see as strong, prestigious and known symbols and brands. First of all, a few facts about the origin of the tourists;

1.1 Destination "Europe"

Most of the Chinese travellers come from the three major economic regions: Beijing and Tianjin, Yangtze River Delta (Shanghai, Jiangsu and Zhejiang) and Pearl River Delta (Guangdong, Hong Kong and Macao); these are the regions that have

---

[1] Stockinger, J. (2008): Marketing in China. In: http://www.austriatourism.com/xxl/_site/int-de/_area/465223/_subArea/465282/_subArea2/479820/_id/485807/index.html.

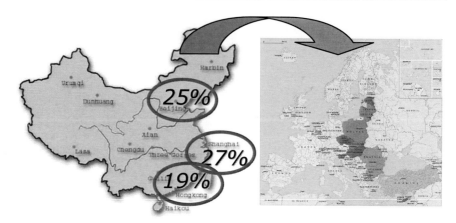

Fig. 1. "Destination Europe": main source markets: Beijing, Hong Kong, Guangzhou, Shanghai[2]. Source: Echtermeyer (2006)

profited most from the economic growth over the past 20 years, so people there can afford to travel abroad. People from Beijing, Guangdong, HKG and Shanghai dominate outbound tourism. 71 percent of travellers come from these regions, which generate most of the nation's GDP and contain more than a third of the entire Chinese population (GNTB, 2005):[3]

- **Shanghai** is the source market of 27% of Chinese incoming tourists to Europe.

- **Beijing** is the source market of 25% of Chinese incoming tourists to Europe.

- **Guangzhou** (formerly known as Canton) constitutes 19% of Chinese incoming tourists to Europe.

The remaining percentage comes from other provinces.

1.2   Destination "Germany"

As far as incoming tourism to Germany is concerned:[4]

52 % of the tourists are from the four large eastern cities: Shanghai (15 million inhabitants), Beijing (13 million), Hong Kong and Guangzhou, capital of the

---

[2]   Echtermeyer, M. (2006): Speech on "Chinese incoming tourism to Europe" on the occasion of the *Internationale Tourismus Börse* (ITB, International Tourism Fair) in Berlin 2006.

[3]   Cf. Deutsche Zentrale für Tourismus (German National Tourist Board, ed., 2005): Marktinformation China, p. 3 – 10, Frankfurt.

[4]   Cf. IPK International (ed., 2004): World Travel Monitor 2003. See also: Deutsche Zentrale für Tourismus (German National Tourist Board / GNTB, ed., 2005): Marktinformation China/Hong Kong, p. 4 – 9, Frankfurt.

Guangdong province (72 million inhabitants). Guangdong is the neighbouring province of Hong Kong and Macau and accounts for 12% of all trips abroad organised by travel agencies.

European destinations, all of which would like to benefit from the growing Chinese inbound tourism, are feeling the growing pressure of competition and innovation. Smart destination managers know that good brand management is the key to success in international markets. This includes specialties, superlatives, new attractions and also traditions. The fight for attention is pushing the limits further and further, according to the mottos of BTO or DTO; i.e. "better than others" or "different to other", hopefully in order to become more well-known than the competition.

Destination marketing is therefore more ostensibly known as the competition of symbols: Karl Marx (Trier) versus Beethoven (Bonn) or Heidiland (Switzerland) versus Don Quijote de la Mancha (Spain). However, the Chinese tourists are not just interested in bare symbols, but also in the **life story of the person whom the symbol represents** – more than the historical background behind it. When they come to Trier, they are less interested in its Roman history but in the

Fig. 2. The "live" story of the Pied Piper of Hamelin played by locals as part of a travel package for Chinese tourists in Germany[5]

---

[5] Echtermeyer, M. (2006): Speech on "Chinese incoming tourism to Europe" at the International Tourism Fair (ITB) in Berlin.

living conditions that Karl Marx experienced in the house in which he was born. The Chinese are very fond of the Brothers Grimm's fairytales; they know the life story of the Austrian empress Sissi, Heidi from the Swiss mountains, the Pied Piper of Hameln and Don Quijote de la Mancha. All of these regions have successfully been taking advantage of their knowledge of Chinese guests' preferences and are already offering them individually-tailored tourist packages (e.g. see Fig. 2).

Destinations with less "personal history," but a lot of prestige, have a magnetic attraction for Chinese visitors, too, e.g. the Hugo Boss Outlet Centre in Metzingen or the famous racing track in Germany, the Nürburg Ring, as well as BMW's brand land (an interactive brand experience on the factory premises) "BMW World" in Munich.

The future belongs to destinations which attract guests with the charisma of strong brands or famous personalities.

> "Tourism is a big branding business. From powerful national governments to local economies, regions across the globe are doing their best to create a brand that attracts tourists and their money."

(Baker, B. 2007)

2   The Prerequisites and Benefits of a Strong (Hotel) Brand

The same rules apply to both brands in the hotel business and brands of other consumer goods industries;

> "A brand is a name, term, sign, symbol, or design, or a combination of these elements that is intended to identify the goods and services of one seller or group of sellers and to differentiate them from those of competition".[6]

Or put differently; brands are crash barriers in the confusing supermarket of messages that assist and guide customers, fulfil their emotional wishes and unburden them. In our complicated world, our culture of abundance ("ZUVIELisation"), customers are looking for preferably easy solutions to problems.

For example; an easy solution for Chinese guests checking into a hotel after a long journey would be to offer them a "pillow menu", allowing each guest to choose their preferred pillow (thick, medium, thin) to suit their individual sleeping needs. This easy and economical solution to the problem goes beyond the core proposition (bed/standard pillow) and adds value for the visitor.

---

[6]   American Marketing Association (published in 1960): Marketing Definitions – a Glossary of Marketing Terms, Chicago.

Particularly in the hotel business, customer loyalty is less dependent on the price per night than on the product's core- and additional benefits. The famous 20:80 rule applies here; most of the time, core benefits make up around 80% of costs but often account for just 20% of the most important impressions left on the customer, whereas additional benefits often only account for about 20% of costs while the crucial impressions left on the customer account for about 80%. Hence, branded hotels stand to gain by offering emotional added value and additional benefits (e.g. individual extras) alongside the rational performance features (clean rooms/beds), leading to guests favouring certain hotels and becoming repeat customers.

Four Points Hotels, for example, positioned themselves in the American market a long time ago using the motto "just like home," offering free coffee and water in all of their hotels coupled with great value for money. Their strategy matches the 20:80 effect as described above, achieving a great brand/image effect at very little cost.

For suppliers, a brand is a promise made to the customer, which at the same time equates an obligation to adhere to the promise of the brand (in line with the core benefits and additional benefits). Suppliers can generally benefit greatly from brand management as the following figure (Fig. 3) indicates.[7]

Creating a brand is not difficult. The challenge is to make it strong and well-known, so that it works in practice and actually gets recognised. There are numerous brands in the hotel business – but only a few have managed to establish themselves worldwide.[8]

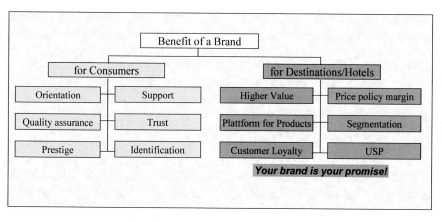

Fig. 3. Benefit of a brand for customers and suppliers. Source: Echtermeyer (2008); cf. Meffert/Buhrmann/Koers 2002

---

[7]   Echtermeyer, M. (2008): Self-designed diagram following Meffert/Buhrmann/Koers (2002).

[8]   Widmann, M. (2008): Wir leben in einer gebrandeten Welt, In: Top Hotel (ed., 2008), No. 9, p. 35.

Once a supplier has established a brand and is better-known than its competitors, it is vital to maintain this advantage and expand on it. In this process, new products should not be developed from the company's perspective but should rather be consumer-orientated, following the motto: "It is not the fisherman that should savour the bait, but the fish".

Brands are most important in the extreme ends of the market; that is to say, in the low budget sector and in the luxury sector, although they have differing functions in each. In the low budget sector, a brand offers safety and the customer avoids any nasty surprises. Low budget brands are usually highly standardised. In the luxury sector, however, brands are seen as status symbols and therefore cannot be uniform. Although the brand guarantees certain basic standards, guests expect different hotels of a chain to be individually designed, and that a Ritz Carlton in Los Angeles looks different to a Ritz Carlton in Shanghai.[9]

The brand name and the brand image should be unmistakable and suit the target group. The brand name and image in turn determine the product and the value of the brand.

In 2005, the American Starwood hotel chain managed to give each of its seven hotel brands a highly distinctive look by radically adjusting its branding strategy. Initially, for example, they fitted the rooms of the hotels of the Westin brand with four-poster beds and then marketed them with their "Heavenly Bed" campaign. Other hotels quickly followed suit and copied the idea, trying to outdo each other. Thus, the original idea and USP were no longer distinctive. This concept worked better later, when they used new, indistinguishable ideas and identities, and co-branding in partnership with other enterprises. True to the new motto, "Recharging in the Westin", the hotel chain started a wellness and fitness promotion in conjunction with Reebok, and customers could also take advantage of fully-tanked BMW rental cars. This led to an overall positive image transfer ("spill over effect") and strengthened the Westin brand. As Kotler (2005) and James (2006) quite accurately describe; "Co-branding partners seek to mutually enhance each other's service or product brand through close association".[10] With a co-branded product, the consumer combines the different sets of brand attitudes towards the participating brands to form one composite concept.[11]

Making a brand stick in people's minds is more than just a matter of having a logo or a flag on the roof. The essence of a brand needs to be felt all the way

---

[9]   Ibid.

[10] Kotler, P./Wong, V./Saunders, J/Armstrong, G. (ed., 2005): Principles of Marketing, 4th ed., p. 113, New Jersey.

[11] James, D. O./ Lyman, M./Foreman, S. K. (2006): Does the tail wag the dog? Brand personality in brand alliance evaluation, In: Journal of Product & Brand Management (published in 2006), 15 (3), p. 173 – 183.

through from the initial contact to the customer to the last, and is mainly supported by the objective core and additional benefits of the product. The successful implementation of a brand into customers' minds requires the following:

- a clear identity (i.e. identifying features in and out),

- striking profile (a distinct corporate identity),

- high value "Story telling" (in which a brand delivers a clear message),

- brand living (all stakeholders of the brand live according to the values expressed by it),

- "lighthouse function" (i.e. brand-giants that can broadcast their brand's message through their sheer size, and create added value for their customers).[12]

According to a press release by DEHOGA (Deutscher Hotel und Gaststätten Verband / German hotel and restaurant association) in March 2008, the **branded hotel business** remains on the road to success with a market share (in terms of sales) of 55.4 % in Germany.[13] However, international surveys on brand awareness show that hotel brands are still mostly at the bottom of the pile. Few have achieved global brand recognition. The importance and value of a strong, well-managed brand becomes clear in light of the following rankings of international brands according to their value (Table 1)[14].

Table 1. Brand values of major international hospitality businesses

Hospitality Business	Brand Value in Billion US$
Starbucks	$ 41.8
Disney	$ 32.6
Mc'Donalds	$ 25.3
Burger King	$ 2.4
Hilton	$ 1.2

Source: Kotler et al. (2006)

---

[12] Cf. Reiter, A. (2008) and www.ztb-zukunft.com.

[13] IHA (ed., 2008): Pressemitteilung ITB. (press release of the ITB) – p. 2.

[14] Kotler, P./ Bowen, J.T./Makens, J.C. (ed., 2006): Marketing for Hospitality and Tourism, New Jersey.

3 Prerequisites, Challenges and Benefits of a Strong Destination Brand

The same principles do not apply to brands of destinations as to those of the hotel industry or other consumer-goods industries. Ritchie und Crouch (2003) define destination brands as follows:

> "A destination brand is a name, symbol, logo, work mark or other graphic that both identifies and differentiates the destination; furthermore it conveys the promise of a memorable travel experience that is uniquely associated with the destination; it also serves to consolidate and reinforce the recollection of pleasurable memories of the destination experience". [15]

Positioning tourist **destinations** in terms of **emotional dimensions** is better in the majority of cases than concentrating on the objective core benefits, as the ranges of products and services offered by different regions nowadays are often very similar. The specific emotional benefit must be expressed by its positioning, thereby differentiating a destination from its competition. Knowledge of the region's core competences is thereby crucial, for example:

The Austrian national tourist office, *Österreich Werbung* (ÖW), introduced a paradigm shift a few years ago and changed the product-orientated marketing strategy to one creating a sense of identity. ÖW was the first European NTO to promote **values** rather than themes. Austria positioned itself as the "world's most charming holiday destination". They implemented these values through "story telling," using two "charming penguins" called Joe und Sally. However, using two penguins as mascots for Austria was not as successful as expected. Particularly since there is not a single wild penguin in Austria, and tourists certainly do not associate Austria with penguins.

The website was re-designed – no longer with penguins – and is successful in comparison to other European destination websites. The ÖW China invests large amounts of money into search engine marketing in the Chinese web each year, and expects to receive the following number of hits on its Chinese website in 2009:[16]

Visitors: 120.000
Page views: 750.000

Theoretically, Austria can reach up to 253 million potential Chinese internet users with its link www.aodili.info. Furthermore, TV and radio advertising, which many

---

[15] Ritchie, J.R.B./Crouch, G.I. (ed., 2003): The competitive destination, a sustainable tourism perspective, p. 165.

[16] Neumüller, P. (2008): Ihr Draht zu 253 Millionen Internetusern. In: http://www.austria-tourism.com/xxl/_site/int-de/_area/465217/_subArea/index.html.

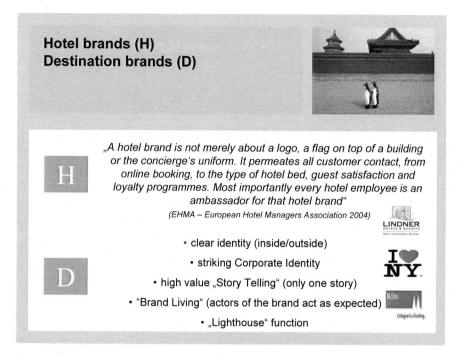

Fig. 4. Hotel brands versus destination brands. Source: Reiter, A. (2008)/www.ztb-zukunft. com

European destinations still neglect to make use of, are of major importance in a country with a population of 1.3 billion. The top right hand corner of Fig. 4 shows the Austrian advertisement with the two penguins, below which there is a short summary of important aspects of hotel and destination brands.

In comparison to products or other services, the budget destination marketers have at their disposal is considerably lower. Therefore destination marketers have to make the most of this limited budget and carefully choose how and where they want to invest it. "They have to outsmart rather than outspend the competition – and that means creating innovative, attention grabbing advertising on a budget and maximizing the amount spent on media".[17]

Hudson (2008) divides the main challenges of destination branding into five different categories; limited budgets, politics, external environment, destination product and creating differentiation.[18]

In addition to limited budgets, politics plays a major role in the public sector of destination branding. Destination marketers have to reconcile a range of local and

---

[17] Hudson, S. (2008): Tourism and Hospitality Marketing – a global perspective, p. 402, London.

[18] Cf. Ibid. p. 402–404.

regional interests and promote an identity acceptable to a range of audiences. Therefore, the challenge is to achieve a balance between the conceptions of political stakeholders and those of the destination management/marketing organization (DMO). Even political entities from other countries can have a substantial influence on destination branding activities.

An example for such a case is the edgy global destination marketing campaign called "Where the bloody hell are you?" of Tourism Australia. When it was launched in 2006, countries like Canada, Singapore and even Britain refused to broadcast the television commercial for to several different reasons. One criticism was that a man was showed taking a sip of beer, whereas others were concerned about the expression "bloody hell".[19] However, this did not result in a total marketing disaster for Australia, but in an overwhelming success in terms of media coverage and public attention.

This again shows that successful brands manage to reach consumers on an emotional level, because emotions are a decisive factor in the destination selection process. In relation to this, Morgan and Pritchard (2000) state that: "to successfully create an emotional attachment a destination brand has to be credible, deliverable, differentiating, conveying powerful ideas, enthusing or trade partners and resonating with the consumers." [20]

The process of destination brand building is normally considered as a five stage process which is displayed below (cf. Fig. 5).

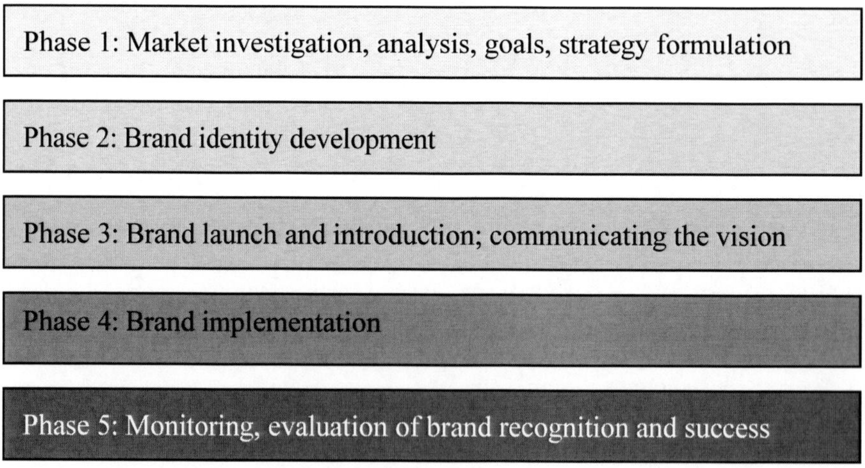

Phase 1: Market investigation, analysis, goals, strategy formulation

Phase 2: Brand identity development

Phase 3: Brand launch and introduction; communicating the vision

Phase 4: Brand implementation

Phase 5: Monitoring, evaluation of brand recognition and success

Fig. 5. The five phases in destination brand building[21]

---

[19] Hudson, S. (2008): Tourism and Hospitality Marketing – a global perspective, p. 348–351, London.

[20] Morgan, N./Pritchard, A. (Ed., 2000): Advertising in tourism and leisure, p. 281, Oxford.

[21] Morgan, N./Pritchard, A. (Ed., 2000): Advertising in tourism and leisure, p. 69–71, Oxford.

In order to identify a destination's core values, market research has to be conducted. This research should investigate which unique attributes and characteristics consumers associate with the destination. As previously mentioned, branding is about stressing the uniqueness of the destination, which is why the consumers' perceptions and cultural backgrounds are very important in this context. In addition, the key competing destinations should be evaluated in terms of attributes and features.

Important questions in this context are;

- What are the tangible, verifiable, objective, measurable characteristics of this destination?

- What benefits to the tourist result from this destination's features?

- What kind of psychological rewards or emotional benefits do tourists receive by visiting this destination?

- What does value mean for the typical repeat-visitor?

- What is the essential nature and character of the destination brand?

From time-to-time it might be necessary to modify some brand characteristics. The secret is to "evolve continually and enrich the original brand personality, building on the initial strengths to increase their appeal and broaden the market".[22]

4   Cultural and Economic Reasons for Chinese Tourists' Brand Consumption

Brands are an area of conflict between globalised economic values and non-globalised diverse cultural aspects. As a result, brand awareness has varying characteristics, dependent on culture.

Experts talk about the existence of a clear East-West divide[23]: Eastern Europe has been affected by the extended absence of western branded goods. This is why well-established and distinguished (western) brands, which embody personal status, enjoy great significance. This also applies to hotel brands and will remain the case for some time.

Even people in Asian countries like Japan or China are highly brand aware and have a pronounced class and hierarchy consciousness, owing to their collectivist-influenced social order. The strong brand awareness is present in all branches – from fashion to cars, from hotels to 'must-see' destinations like Paris, Munich and the

---

[22] Hudson, S. (2008): Tourism and Hospitality Marketing – a global perspective, p. 406, London.

[23] Cf. Wissmath, R. (2008): Identifikation, Individualität & Internet, In: Top Hotel (ed., 2008), Nr. 9, p. 34 and Widmann, M. (2008), ibidem, p. 35.

nearby castles, and Rome. Chinese tourists look for status symbols. It is more prestigious for them to buy an original Hugo Boss in Germany, an original knife in Switzerland and perfume in Paris than to acquire them in a shopping centre in Beijing. In Chinese culture, giving branded goods as presents is called 'giving face'.

However, it is impossible to speak generally and sweepingly of 'Chinese brand preferences'. China is a vast land with 1.3 billion people, whose local culture, industrial structure and consumer buying habits vary considerably from one province to the next. European tourism managers in marketing should therefore not speak about 'Chinese tourists' in general, but should develop an understanding of the varying interests and desires of tourists from the particular provinces. The source markets Beijing, Shanghai and Guangzhou are the three cities with the most accurate characteristics of the Chinese outbound market. More information on this can be found in a study by the World Tourism Organization 2003, titled 'Marketing on Chinese outbound tourism.'[24]

5   Practical Examples

The following text shows extracts from the results of a round of talks on the occasion of the International Tourist Fair (ITB) in Berlin 2007 on the topic *"Destination Brand – brands stimulating tourism growth"* : Present at the round of talks were **Wolfgang Bauer (Chief Executive Officer Holy AG / Outletcity Metzingen), Dr. Walter Kafitz (Chief Executive Officer Nürburgring GmbH) and Marcel Schneider (Chief Executive Officer TUI China Travel Co. Ltd.), as well as Prof. Dr. Monika Echtermeyer (mediation).**

All the experts agreed that strong brands and brand lands are very important for tourism, in particular for the Chinese but also generally for the international market. Emotional benefits of a brand (feelings of prestige, happiness) are predominant in the choice of destination.

A prominent example is the development of the small German city of Metzingen, the domicile of the German clothing producer Hugo Boss AG which was de-

---

[24] "People from Beijing stress experience in life and cultural taste as well as personal participation. Shanghai people are very practical. They will calculate how much they spend and how many attractions they will see. People in Guangdong pay much attention to enjoyment, such as good food and entertainment. When a new destination is opened, the market in Beijing and Guangdong will immediately get moving. But people in Shanghai will wait and see before they take any action. People in the three cities have different degree of fondness for the same destination. For instance, passengers from Guangdong like Japan more than passengers from Beijing and Shanghai do. Winter tours to Korea sell much better in South China than in North China. Such differences also apply to a city. For example, passengers from Beijing and Guangdong love Cairns, Australia, much more than the people in Shanghai do" (Source: WTO: "Report on Shanghai Society 2001", p. 130).

veloped from a small factory outlet into the internationally famous 'Outletcity' of Metzingen, with visitors from 165 nations last year and more than 2.5 million visitors annually. With the targeted acquisition of international top-brands (Escada, Burberry, Bally etc.), tourists now come to Metzingen from all over the world in order to shop at bargain prices, and at the same time have an emotional shopping experience in places such as an old factory building where Hugo Boss genuinely is based. Metzingen is one of Germany's top cities in the Global Refund shopping success barometer.

Dr. Walter Kafitz (Chief Executive Officer Nürburgring Ltd.) declared that the Nürburgring brand has been famous far beyond German borders for many years (20% of the 2 million visitors are foreign guests), however the establishment of the brand among Chinese guests and the tourist boom associated with them is, on an international scale of comparison, still in its very beginnings. As president of the international association of track owners *("Streckenbesitzvereinigung")*, Dr. Kafitz is currently in the process of cooperating with the Shanghai region and the race track in Shanghai, with the goal of attracting more Chinese tourists to the Nürburgring. In the minds of Chinese motorsport fans, the Nürburgring is already a well-established brand. The Nürburgring (also known as "Green Hell") is the most famous race track in the world. Furthermore, the Nürburgring is also a destination – a brand destination – that people simply want to visit.

According to information from Marcel Schneider (Chief Executive Officer TUI China Travel Co. Ltd.), a trip to Europe including buying 'the right brand items' is of great significance for a Chinese tourist. Tourists come back with branded goods, because they carry prestige, provide them with 'face' and can even give face to others (relatives, friends and colleagues). However, the reasons for the purchase of brands are not only cultural but also economic: the currency exchange in China is legally limited to 3000 US $ per journey abroad. Expenditures on credit cards are however unlimited. Due to luxury taxes, which are levied by the Chinese government on certain luxury goods, Chinese citizens have to pay 35% in taxes.

This is very much to the advantage of Wolfgang Bauer of Outletcity Metzingen. Chinese guests spend a lot of money on top-quality branded clothing in Metzingen and accordingly feel happy to be shopping without the luxury tax. The following illustration (Fig. 6) shows the most popular hotel – and commercial brands of Chinese tourists in Germany.

According to Dr. Walter Kafitz, even the Nürburgring GmbH as a service enterprise sells feelings of happiness when they take Chinese delegations for a drive along the longest and most challenging race track in the world. Particularly demanding shopping/brand enthusiasts have the opportunity to buy a Nürburg-edition Aston Martin for 225 000 €, or a Nürburg-edition mobile phone for 6 200 €. Even for the more reserved spenders there is a range of merchandise including souvenirs and gifts bearing the destination-brand "Nürburgring".

TUI China Travel Co. Ltd. is a worldwide quality brand and should also be recognized as such in China. The company is therefore focussing its efforts exclusively on delegation business. This is partly made up of high-ranking delegations

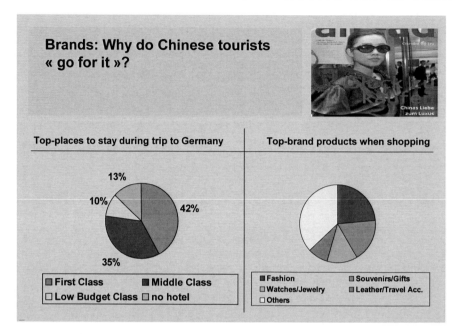

Fig. 6. Top hotel categories and top product brands of Chinese tourists in Germany. Source: Echtermeyer, M. (2008)/Global Refund (2007)

from provincial governments or companies that want to combine a holiday around Europe with a business trip. The aspect of education, of seeing, of discovering the business side is one part of it, but there is also the experience, seeing, feeling, and shopping on the leisure side. TUI China is clearly aiming at the strata of society in China with high buying-power. At the moment, TUI China is focussing on the delegation business for licensing reasons, which allows the company to generate relatively high margins in its operational business in comparison to volume business, which is extremely price-sensitive in China, even though it is still very young. The USP of the "TUI China" brand is, according to information provided by Marcel Schneider, very clearly; quality. When, for instance, Chinese delegations travel to Europe or other parts of the world, they are, wherever possible, taken care of solely by TUI companies. This ensures the delegations' security as to the quality of service they receive. For, in comparison to mass tourism, such high-ranking delegation travel has a lot of prestige and "face" attached and the guests attach a great deal of importance on everything running smoothly. By now, TUI has a worldwide network of inbound agencies with regulated quality standards, which is a novelty in Chinese outbound travel.

With regard to the choice of brands and quality, there is a difference between first-time visitors and second- and third-time visitors. The first-time visitor (mostly a package-holiday tourist) is restricted by price. The second- and third-time visitors are slightly sobered after the initial travel experience and want to be protected from certain mistakes and discomforts. They are already more quality-conscious and favour international brands and hotel groups with 3-5 stars. The differences between these target groups in demand behaviour and in the choice of brands are striking.

In answer to the question of future core/additional experience of brands, Dr. Walter Kafitz (Chief Executive Officer Nürburgring GmbH) says; "We are currently developing ourselves into a year-round leisure and business centre. We are investing 250 million Euros. There will be more hotels, on the low-budget side there will be a so-called motorsport-resort, bungalow residences, there will be themed catering and shops, so that it will become a fully-fledged destination where people stay overnight and even stay for extended periods, and so are prepared to spend more money. We want to increase our attendance figures from 2 million to 4.1 million visitors. Construction is currently in progress and the opening is set for early 2009. We are developing a boulevard directly next to the Nürburgring, a promenade, roofed, with shops, with an arena, with a convention hall, with an event hall, with a 4-star superior hotel, with a 3-star hotel, with themed catering, a karaoke bar and steakhouse, a country bar and a motorsport resort. We are handling this investment of 250 million Euros together with private investors."

Wolfgang Bauer (Chief Executive Officer Holy AG/Outletcity Metzingen) is also dealing with the offer of additional experience. A catering concept to meet international demands is soon to be implemented, in order to enrich the core shopping experience with added value. "We are also starting to look into the hotel industry as, with 2.5m guests and 213 hotel rooms, we are of course acutely dependent on the region. We are working intensively on the integration of the whole area of interesting hotels and tourist features in the region. We are developing a concept that will capture additional emotional value, and which will prepare the outlet stores for the expectations, imaginations and desires of the various nations. We want to offer our guests additional value that they can't get in other destinations.'

In conclusion, one could say that the development of a brand image is highly important in future competition, whether it be in the Chinese, Indian or Russian markets, or any other for that matter. It is of particular relevance for the first-time customers. Every country has particular products, symbols or characteristics that define the brand image of the country and as such act as export produce in the tourism industry. Successful tourism offers are always simple solutions in our society of abundance. They recognise the slightest conscious and unconscious desires of the consumer; they gear themselves according to intangibles, i.e. untouchable things such as brand value, needs and wants of the customer, and integrate them into attractive products. The key to becoming a successful brand-company or destination should be identifying the future market potential, and continually converting this into innovative intangibles.

Future: How does this « hunger for brands » change the destination landscape ?

„Tomorrows tourism marketing environment will be one defined by ever-increasing competition, greater product variety and consumer choice in which **INTANGIBLES** *such as* **brand values, experiences, emotional benefits** *and celebrity will be the key differentiating factors."*

(Morgan, N./Pritchard, A. 2007)

Fig. 7. Key differentiating factors in tomorrows tourism marketing

References

Stockinger, J. (2008): Marketing in China. At: http://www.austriatourism.com/xxl/_site/int-de/_area/465223/_subArea/465282/_subArea2/479820/_id/485807/index.html

Deutsche Zentrale für Tourismus (ed., 2005): Marktinformation China, p. 3-10, Frankfurt

IPK International (ed., 2004): World Travel Monitor 2003, und Deutsche Zentrale für Tourismus (ed., 2005): Marktinformation China/Hong Kong, p. 4-9, Frankfurt

Echtermeyer, M. (2006): Speech: „Chinese Incoming Tourism to Europe" on the occasion of the ITB 2006

American Marketing Association (ed., 1960): Marketing Definitions – a Glossary of Marketing Terms, Chicago

Baumgarth, C. (ed., 2001): Markenpolitik (Brandmanagement), p. 6, Wiesbaden

Echtermeyer, M. (2008): Self-designed diagram following Meffert/Buhrmann/Koers (2002)

Widmann, M. (2008): Wir leben in einer gebrandeten Welt, In: Top Hotel (ed., 2008), Nr. 9, p. 35

Kotler, P./Wong, V./Saunders, J/Armstrong, G. (ed., 2005): Principles of Marketing, 4th ed., p. 113, New Jersey

James, D. O./ Lyman, M./Foreman, S. K. (2006): Does the tail wag the dog? Brand personality in brand alliance evaluation, In: Journal of Product & Brand Management (ed., 2006), 15 (3), p. 173-183

Reiter, A. (2008) in: www.ztb-zukunft.com

IHA (ed., 2008): Press release ITB, p. 2

Kotler, P./ Bowen, J.T./Makens, J.C. (ed., 2006): Marketing for Hospitality and Tourism, New Jersey

Ritchie,J.R.B./Crouch, G.I. (ed., 2003): The competitive destination, a sustainable tourism perspective, p. 165

Neumüller, P. (2008): Ihr Draht zu 253 Millionen Internetusern. In: http://www.austria tourism. com/xxl/_site/int-de/_area/465217/_subArea/index.html

Hudson, S. (2008): Tourism and Hospitality Marketing – a global perspective, p. 348-351, 402-406, London

Morgan, N./Pritchard, A. (ed., 2000): Advertising in tourism and leisure, p. 69-71, 281, Oxford

Wissmath, R. (2008): Identifikation, Individualität & Internet, In: Top Hotel (ed., 2008), Nr. 9, p. 34 and Widmann, M. (2008), ibidem, p. 35

WTO (ed., 2002): "Report on Shanghai Society 2001", p. 130, Madrid

Luxury Tourism – Insights into an Underserved Market Segment

Klaus-Dieter Koch

1   Brand Building in the Post Advertising Era

Human beings want to trust. Trust provides a sense of assurance and relief. Trust simplifies as well as accelerates the decision-making process and builds confidence. In an increasingly complex world, confidence in decision-making and purchasing is a precious asset and can, at times, be more important than stimulating products or services themselves. In fact, the additional value that a brand name offers is assurance and relief through trust. Yet, trust and assurance are not "the initial stage of everything" as claimed by an advertising campaign. They are rather the final stage of a very long series of gestures of confidence. Therefore, developing a brand requires more time than generally assumed.

Indeed, trust-building requires different amounts of time depending on corporate history and the sector one is active in. During this time, the product or service has to maintain high standards of quality. Quality standards should exceed the customer's expectations of the brand. This is considered to be a basic rule of business and cannot be ignored; not even by modern marketing methods which often promise a great deal more than they can live up to. So, what is the rule nowadays? Companies exaggerate hopelessly and promise more than their products are capable of. The situation is even more dramatic in the service industry: it is the case of banks, insurance and telecommunication companies or travel agents. All promise the moon with the help of gigantic PR campaigns but cannot, or can only just, keep their promises. As each of their offers bears a company logo, managers believe they are actually developing a brand, when really, their marketing methods are nipping in the bud any effect a brand name might have. How does the consumer react to all this? If he/she does not ignore and disregard the message in the first place, but in fact accepts one of the highly praised offers only to see his/her expectations disappointed, then the company logo in question will immediately and directly be linked to the disappointment and stored as a whole in the consumer's memory – sometimes for decades if the degree of the disappointment is particularly strong. This leads to the build up of "negative prejudices" which do not necessarily follow immediately on the negative experience, but often emerge later on, when the experience lies so far back in the past that one is hardly able to recall the reason why this company logo evokes negative feelings and associations.

In order to avoid overburdening ourselves, negative memories and broken promises are stored in combination with the company logo. Everything is reduced to this one element which clearly says: "Hands off!" This is what we store in our minds and then remember and this is all it takes for survival on the consumer market.

Therefore, when developing a brand, the golden rule is:

1.1   Quality Comes First, PR Activity Comes After

At some point market analysts detect this situation and start talking about negative brand image. What marketing steps are taken then? Exactly! An image campaign in launched along the lines of: We are not what you think we are. We are quite different. This image campaign collides with the negative image stored in one's memory. What happens next? Nothing. Then, who should we trust more? Our experience which time has downgraded to prejudice or the advertisement which caused the negative experience in the first place? So:

1.2   Prejudices Cannot Be Tackled Through Advertising

Well then how? Simply through quality! Long-lasting top quality together with the positive experience thereto connected are stored in the consumer's memory in much the same manner as happens with negative experience. At some point in time this type of experience also detaches itself from judgements. This is what we call "positive prejudices". The latter have the benefit of being as persistent as their "negative" siblings.

The advantages are obvious. Without positive prejudices, purchasing decisions, such as travelling decisions, would be too risky. After all, it is about deciding on how to spend the best weeks of the year, or how to make the next business trip as safe, efficient and pleasant as possible. In a city one is not familiar with, the "taxi" sign on vehicles waiting outside the airport terminal building, is in itself a sign of trustworthiness. Major hotel chains and their brands provide orientation and one should more or less know what to expect from them. Finding an airline for a connecting flight would almost be impossible without the help of a brand that entails a series of positive prejudices. Without prejudices consumers would have to compare their wishes and expectations with what is on offer each time, to be sure they are getting what they want.

In our modern, information-flooded society, the main task of brand-names is to turn long-lasting, top-quality-induced positive prejudices into solid monuments of trustworthiness.

For this simplification, assurance and relief, consumers are potentially willing to pay far more than for a comparable product or service that does not offer the added value of being a brand. However, not only this: consumers do not only pay a surcharge, they also prove loyal for an above-average longer period of time. The risk of changing is simply too high even though cheaper offers are appealing. After all, experience has taught one to know better. The third advantage, which can be of added value to brand-owners, is the increase in the brand cross-selling. Trust does not only create consumer loyalty; it also enhances the attractiveness of other products and services within the same brand. The brand can be expanded within its range of credibility. New products can then be introduced at a lower risk. Consumers' enhanced perception of brand messages allow for brand advertisement to stand out from the rest belonging to the same product category. Relevance is a prerequisite for perception and is gained by achieving top quality. In this way, the brand name helps marketing regain its effectiveness.

Last but not least, the most crucial advantage: word-of-mouth recommendations increase, which implies that it is the customers who take over the marketing activity – and they do so with an impartiality and power of persuasion which even the best marketing strategies cannot possibly achieve. This considerably lowers the cost of customer acquisition. According to a study carried out by Mediaedge, when deciding on a purchase, approx. 75% of consumers trust product recommendations made by family and friends.

Although this may sound excessively banal to some, providing consumers with a sense of assurance and relief through security and simplicity can be an item of added value to the brand. Yet, these elements do not only matter when consumers are standing in front of supermarket shelves and are about to decide what to purchase.

Keyword Social Recognition: When school children demand a specific pair of sneakers, it is more about survival in school and among friends rather than about fashion or functional issues. Contrary to repeatedly-made claims, such behaviour is not instilled into children by the producers of branded articles, it is simply what human beings are all about. It is vital for survival in any society throughout the world to express one's identity and hierarchical position through the clothes one wears and the items one carries around. Clothing is to managers, politicians, wives of oligarchs or boy scouts what elaborate body painting is to the indigenous people in Papua New Guinea or what insignia are to noblemen or clergymen. They are encoded and simplified messages in much the same way as colours (politicians), order (church), badges (boy scouts), patterns (noblemen) and plastic surgery (wives of oligarchs) are to church, politicians, boy scouts, noblemen and wives of oligarchs respectively and which publicly document the social position, the power, the achievements, the level of responsibility of the bearer thus coordinating essential co-existence within the group as well as allowing for transparency and order. Ever since the development of the traditional branded article in the mid-19th Century, these highly dense message systems have been employed to show one's circle of acquaintances – and thanks to greater mobility to strangers alike – who one is and what one has achieved.

Keyword Self-definition: Greater globalization-fuelled individualism calls for increased orientation. Who do I want to be today and in the future? Well-structured brands, with their clear-cut value system, help consumers transfer, define and find an archetype which suits them. These cultural codes bring orientation, inspiration and stability to the development of an individual identity.

Keyword Inspiration: Brands can distinguish themselves thanks to their underlying value system and can, therefore, cause exciting contrasts. In cooking, the combination of Emmental cheese and Valrhona chocolate creates an entirely new taste- as well as flavour-experience. In fashion, Levis jeans are designed with Swarovski crystals and, within the world of design and art, superstar Takashi Murakami works together with Louis Vuitton. Brands do not only inspire the owner and the industry to strive for new and exciting results but, more importantly, they also inspire consumers. Nobody combines standard with standard. Contrasts are necessary and contrasts are exactly what brands provide. Unlike any other system, brands stimulate the creativity of their customers. However, clear-cut brand characters, which know exactly what they stand for and what they do *not* stand for, are a precondition for this.

Keyword Individuality: First of all brands create individualism. This individualism is the driving force behind consumption. The great possibilities lead one to desire more and new things, which is what keeps market economies alive. A market segment can be tapped and better exploited, if existing brands are positioned in a clearly distinguishable manner. If this is not so, the price prevails and sales plummet.

Keyword Human Needs: Wishes, dreams, hopes and desires describe the emotional hunger of human beings. To some extent, brands can satiate this hunger and provide a sense of fulfilment. One topic that clearly demonstrates this is collecting as a hobby. While talking to a passionate collector, one will soon realize how important brands are; regardless of whether it is china, watches, art, vintage fashion, designer furniture, model railways or even beer mats and crown caps. Without brands, many fields of interest for collectors would not exist.

Brand-names stand for assurance, simple and confident decision-making and provide an additional value which goes beyond the product itself. As a result, brands can add a surcharge which by far exceeds the costs of production and materials. Generating preferably two-digit sales and growth figures is what working for and with a brand is all about. Smaller sales and growth figures can be achieved even without an effective brand name.

1.3   Understanding Brands as Living and Breathing Systems

The main goal is to establish a relationship between product or service and consumers.

Selective perception keeps us from drowning in the flood of visual stimuli (e.g. daily exposure to approx. 3,000 brand names). Our brain filters what it considers unimportant thus relieving us from any unnecessary burden. Anything deemed unimportant for life, is considered irrelevant, i.e. it is something one cannot establish a relationship with. The key to perception is establishing a personal relationship.

Long-lasting personal ties can only be built with something that is alive, meaning something that evolves, contradicts and changes. Only then, will we include it in our "relevant set" and look upon the world with new light. This is not the case when it comes to "dead" products. Yet, to make "dead" products relatable again, they need a name, a character, a story, an origin and a distinctive identity. To be precise: they have to be developed into characteristic and vivid brands in order to be perceived.

Living systems also have another interesting feature: they want to grow. Anything that lives, can also grow. It is an impulse. The question is how to direct it?

The neurologists Scheier and Held wrote in their book "Wie Wirkung wirkt" (What effects effects have): "Advertising is supposed to improve the ranking of something within the mind of the consumer." An improvement in ranking is therefore the most crucial criterion for success in marketing and advertising. The idea of a "relevant set" is the basis for many instruments that measure the success of advertising campaigns. In the end, these instruments only intend to detect the ranking of a certain brand or product on the consumer's mental shopping list. The idea of a mental shopping list is however incorrect! There are only two positions in the consumer's mind: first and what comes after that. Studies in Neuroscience have revealed that cortical relief only takes place in the presence of a favourite brand, i.e. the number 1...".

Since the early stages of the modern market and the beginnings of industrialisation, only one formula existed: "low prices lead to greater sales which lead to lower costs which, in turn, mean even lower prices." This formula no longer works nowadays because there are no mass markets left. Only a created illusion of mass markets exists by pooling niche markets. Additionally, even though most markets are saturated, most companies and managers still believe in price competition. For many years now, they have been reducing costs and producing excessively. In reality, excessive production causes prices to fall and creates a feeling of loss of value in the consumer's mind. Consumers become indifferent and grow weary of consumption. Product excess leads to an even greater pressure on the manufacturing industry and results in another round of price reductions. The sales-boosting effect of this price reduction is now even lower.

The only chance to break this vicious cycle is to protect oneself from this price and value collapse by being unique as well as offering products and services that cannot be evaluated or compared in any neutral way. Obviously, it is impossible to meet this challenge just by thinking of each product individually. Their range of benefits are mostly limited and cannot be truly and permanently unique.

Brands are the most effective management tool in global and saturated markets, if one wishes to increase the prices and EBIT. No other system can store high quality like a battery and release the "performance and brand energy" once it is fully charged. The performance energy which has been converted into brand energy can be released immediately whenever necessary and allows the brand to enter a price-performance competition.

2   The Opportunities New Luxury Offers for Brand Management in Tourism

Luxury is exiting. Not only because in a world in which its consumption is undergoing an increasing democratisation it causes tension and individualism, but especially because it anticipates trends and attitudes, which shape mass markets partly with considerable delay.

At the same time, luxury puts pressure on its affinitive elites, both on the creative/productive and consumer side, to continuously create and therefore purchase novelties that allow for individualism. What is considered as luxury today is already mainstream tomorrow. Just think of mobile phones, air travel or bio-foods.

Since luxury is a powerful gauge of future mass trends, both producers and service suppliers, who focus primarily on volumes, should not underestimate the trends and developments of the luxury market segment, if they wish to keep abreast of it in the near future.

What Does the Future Hold for Luxury and Consumption? How does the attitude towards luxury evolve in this age of change in values? This is the starting point of our study "New Luxury & Brands" for which 900 people with above-average incomes were surveyed in Germany, Austria and Switzerland.

2.1   Traditional Prestige Consumption Is Dropping, Self-Realization Winning

The time when people bought a new hand-bag because their best friend did and a new car because of their neighbour did is coming to an end. In prosperous societies people tend to consume more in order to contribute to the fulfilment of their own life expectations. Luxury brands will of course always be used to impress others, with the only difference that the people one wishes to impress will have to have more background knowledge. The reason for this is that luxury expresses itself with ever increasing subtlety. Not only are logos becoming increasingly codified (e.g. the white Apple head-phone cable) or in the case of Hermes or Valextra are they being almost totally omitted, but, as in the case of coffee and chocolate, luxury once again involves market segments that appear to have degenerated into faceless mass markets.

2.2   The Global Management of Luxury Brands Has to Change

In the future, the homogeneous international marketing of luxury brands could decrease. The more the luxury market grows, the stronger the need for differentiated brand management. On developed and prosperous markets, used to dealing with luxury products, luxury is increasingly being dematerialized and considered as an integral whole. Subtle luxury is acquiring more and more importance, while shrill, over-sized logos are frowned upon. On the new Eastern European and Asian luxury markets, consumers still consider luxury brand logos as all-important: they are never large enough. Distinctive expertise on the one the hand, and committed learning on the other, call for an increasingly differentiated form of luxury brand management. Shops, advertisements or events that are identical to each other on a global level will no longer work with the same effectiveness.

Winning luxury brands will not only be more capable compared to their competitors of adapting to the different needs of their global clientèle, but also of honouring different sets of values and orientating their marketing strategies accordingly.

2.3   Luxury Will Continue to Grow

First of all because the luxury market is not affected by recession. Societies are becoming more and more polarized, hence enlarging the group of wealthy people, who represent a reliable basis thanks to their multiple income sources. Since the mid 90s, luxury has grown into a global phenomenon with markets at different stages of development. This immunises globalized luxury brands against fluctuations affecting national economies on a regional level.

Only selective middle class luxury consumers are affected by recession: in this case, as the income basis is not sufficiently developed or revenues mostly come from one single source, such consumers start saving money as soon as energy prices rise or their jobs are at stake. This is when luxury suddenly goes back to being what it once was, i.e. superfluous.

2.4   The Luxury of the Future Encourages People to Develop and Deepens Expertise in Nine Selected Fields

Self-fulfilment consumption: The aim of consumption is no longer to simply satisfy basic human needs, but mainly to fulfil the yearning for self-fulfilment. It is all about improving the quality and intensity of life, having more leisure time and being physically and mentally fitter and healthier. As, according to the people in the surveyed countries, travel and holidays are still listed among the 3 traditional top luxuries (together with house and health), great opportunities open up for those tourism brands that wish to be consistent in the uptake of this new field of self-fulfilment consumption.

Radical Chic: ethical, responsible leisure consumption, based on fair-trade brands and technologies, opens up new fields that allow for effective brand positioning. Diesel technology for vehicles only proved successful after the introduction of particularly high-torque TDI engines made driving fuel-efficient diesel vehicles enjoyable. Caviar, the traditional luxury product, will end up being the least desirable, unless caviar producers finally understand that their ethical responsibilities are, to say the least, as important as the consumption of caviar itself. The reason why bio-food is so successful is that producers have learned that comparatively expensive food cannot be sold only by appealing to reason.

An ethically acceptable attitude is developing even in tourism and people do no longer shy away talking about it. This debate covers all aspects, the journey to a destination, the stay and the consumer's behaviour. Environmental concerns will increasingly drive purchasing decisions. Sustainability and respect for the environment within the framework of the expectations linked to destination and accommodation as well as the leisure pursuits, without cutting back on comforts, amenities, rapidity or service, will lead people to trade up and trade down in their travel purchases.

Expertise consumption: People are driven to product consumption to gain a greater insight into certain topics, to go beyond a new impression and a superficial experience. Deeper insights and authenticity are becoming the new standard criteria. Real experiences that reach beyond fake shows and artificiality hence contributing to the enlargement of guests' experience, will characterize the markets of the future, leaving aside the leisurely, didactic approach which is typical of past study travels give way to an intellectual approach embedded in an exciting and entertaining framework.

Leisure consumption: Consumers chose this type of consumption to have fun and make the most of their leisure time, alone or with friends. In this case the most important values are of a hedonistic nature and aim to create experiences that can be talked about with friends. Maximum fun and minimum time are the benchmark.

Personalized premium: This involves consumers who are seeking to be creative and express their individuality. What counts in this case are tailored solutions that reflect the highest degree of autonomy and originality for which consumers are willing to pay more than necessary. Originality is parmount. This market segment is for proven experts who are able to rely on years of established experience within a special field and are prepared to adapt totally to their guests' needs.

Concierge Service: Assurance and relief come out on top in this segment. Anything that helps the guests to recover from stress and pressure thus giving them the chance to take a break, unwind and get back onto their feet is considered of value. This allows for maximum individualism, without boundaries and rules. Assistance and support with any of the client's needs is appealing to guests experiencing the most different situations in life. Families with many children as well as double-income households with no children and Best Agers, who at last wish to travel to those destinations they have dreamed of all their lives, even if they are partly limited by their health conditions.

Sociability consumption: What characterizes this segment is the desire to spend more time with friends and relatives and use this time as an opportunity to learn, play, be creative, make and share new and unique experiences. In this case it is of little importance to the guest whether the company he/she is with are people he/she was already acquainted with before the trip or people met during the trip itself. Crucial to sociability consumption is the enhancement of the quality of experience by making new experiences together.

Inclusivity: Typical of inclusivity is the strong desire to be part of exclusive spheres and, within the prevailing key group, be regarded as the best, the winners. The predominant elements are belonging within and rising above the mass.

Top of the World: This is the typical, materialistic status-driven consumption, which consists in purchasing the best of the best and show others what one can afford. This field of luxury is always the first to emerge on the markets of the new rich and dominates the first phase of prosperity. Global brands, predominantly of European origins, are favoured, as well as anything that contributes to confirming or better still enhancing one's status.

Is Europe One Market or Many? The US Cruise Companies' Segmentation Problem*

Michael Vogel

Abstract

The U.S. cruise companies Carnival Corporation and Royal Caribbean Cruises control two thirds of the global cruise ship capacity. They share the ambition to significantly increase their presence in Europe but they do not share the same strategy. Various Carnival brands are dedicated to specific European markets. Their ship capacities are allocated to these markets, and their prices are set locally. Royal Caribbean's main brands, on the other hand, are global. Their capacities are not market-specific, and their prices are set centrally in Miami and applied uniformly in all markets world-wide. The strategic differences can be reduced to a fundamentally different approach to market segmentation. This paper uses microeconomic modelling to evaluate both companies' strategies with respect to their economic performance given the heterogeneous market conditions in Europe. The results show that Carnival's European strategy is superior. However, Royal Caribbean's recent acquisition of Pullmantur and their plans to create a cruise operator for France seem to be steps in the right direction.

1 Introduction

In 2006, 3.4 million Europeans went on an ocean cruise (Sbarsky, 2007). Although this number is modest compared to the 9.4 million U.S. passengers on cruise ships in the same year (CLIA, 2007), growth has been much stronger in Europe (Ebersold, 2007). The major U.S.-based cruise companies have long ago recognized the great potential of Europe's cruise markets and are increasing their presence by deploying more ships and investing in marketing and sales infrastructure.

---

* Earlier versions of this paper have been presented at the conference "Cruise Shipping Opportunities and Challenges: markets, technologies and local development" on 4–6 October 2007 in Naples/Italy and will be presented at the 11th DGT Colloquium on 30 November 2007 in Luneburg. The present paper is still work in progress.

Given that Carnival Corporation and Royal Caribbean's fleets represent 46 and 22 percent respectively of the world's cruise ship capacities in gross tons (Mäkinen, 2007; figures include ships on order), given also the two companies' financial strength and industry know-how, both of them undoubtedly have the possibilities and means to shape the future of cruising for Europeans. Yet their actual performance in Europe also depends on the strategies they have chosen as size and power alone may not translate into success.

The question this paper attempts to answer is how effective the strategies are which Carnival Corporation and Royal Caribbean Cruises have adopted to compete for European cruise passengers. Of course a quick look at the numbers seems to settle the issue. In 2006, Carnival Corporation generated about four times more revenue with European customers than Royal Caribbean Cruises (see below), while only being twice as large in terms of global revenue. But this comparison is unfair because it neglects the different starting points and histories of the two companies. Moreover, strategy effectiveness can neither be understood nor evaluated by looking at outcomes alone.

The approach taken in this paper is theoretical. Microeconomic models of the two cruise companies will be developed in order to study how their different strategies perform under conditions which might be considered characteristic for Europe: many markets with varying volumes, price structures, and product preferences. The great advantage of this approach is that it allows comparing strategies in their purest form on a like-for-like basis because all factors influencing the companies' economic performance that are unrelated to the strategies in question can be excluded from the models.

The paper starts with a brief presentation of Carnival Corporation and Royal Caribbean Cruises and their presence in Europe, as well as of their respective brand, pricing and capacity strategies. The main part develops microeconomic models of the two cruise companies to investigate their strategies' performance in various scenarios. For each scenario, the formal analysis is followed by a separate discussion of the findings. Most of the technical bits have been relegated to the Appendix. The paper concludes with a reflection on the models' results and their significance with respect to the recent strategic moves of Carnival Corporation and Royal Caribbean.

1.1   Company Profiles and Presence in Europe

For the financial year 2006, Carnival Corporation disposed of a fleet of 81 cruise ships with a capacity of 143,676 berths and reported revenues of U.S.$ 11.8 billion, a total of 7.0 million passengers, and assets worth U.S.$ 30.6 billion (Carnival, 2007a). For the same period, Royal Caribbean Cruises operated 34 ships with a capacity of 67,550 berths and reported revenues of U.S.$ 5.2 billion, 3.6 million passengers, and assets of U.S.$ 13.4 billion (Royal Caribbean, 2007a). These figures show that the Carnival Corporation is roughly twice as large as Royal Caribbean Cruises.

Carnival Corporation

Carnival Corporation was created as Carnival Cruise Line in 1972 and renamed in 1994. In 1987 the company went public. The following years were marked by many acquisitions: Holland America Line in 1989, Seabourn Cruise Line in 1992, Costa Cruises in 1997, Cunard Line in 1998. In 2003, Carnival Corporation was combined with P&O Princess, adding Princess Cruises, P&O Cruises, P&O Cruises Australia, AIDA Cruises, Ocean Village, and Swan Hellenic to the brand portfolio.

Carnival Corporation operates mass market and specialist brands, contemporary and luxury brands, U.S., European and Australian brands. Three brands serve the UK market: P&O Cruises, Cunard Line and Ocean Village. The brand Aida Cruises is dedicated to the German market. Costa Cruises, finally, started as an Italian brand, and then became increasingly popular also in France, Spain and Germany. Some of the European brands have long histories, dating back to 1837 (P&O), 1839 (Cunard) and 1854 (Costa). Other brands are younger: Princess Cruises was founded in 1965, Aida Cruises in 1996, and Ocean Village only in 2003.

In 2006, Europe represented 30 percent of Carnival Corporation's revenue (Mathisen & Mathisen, 2007), i.e. approximately U.S.$ 3.5 billion. The company's chairman and CEO Micky Arison emphasized that *"The United Kingdom, Italy and Germany enjoy especially strong cruise demand, and Carnival operates the leading cruise brands in the UK and Continental Europe. We believe Europe has tremendous potential for growth and have ordered 11 ships for our European brands which are scheduled for delivery over the next four years."* (Carnival, 2007a, p. 2)

In addition to the newbuilds mentioned, growth in Europe will be fuelled in Spain by the new multi-ship cruise line Iberocruceros, a joint venture between the majority-holder Carnival Corporation and the Spanish travel company Orizonia Corporación (Carnival, 2007c). For 2010, the share of European revenues in Carnival Corporation's total revenues is expected to reach 40 percent (Mathisen & Mathisen, 2007).

Royal Caribbean Cruises

Royal Caribbean Cruise Line was created in 1968. The company went public in 1993 and merged with Celebrity Cruises in 1997. It adopted the name Royal Caribbean Cruises, henceforth combining the two brands Royal Caribbean International (RCI) and Celebrity Cruises. In 2002, Royal Caribbean Cruises and First Choice Holidays jointly created Island Cruises in the UK. In 2006, Royal Caribbean Cruises acquired the Spanish tour and cruise operator Pullmatur, followed by the launch of Azamara Cruises in 2007.

Relative to Carnival Corporation, Royal Caribbean Cruises' business in European markets in 2006 was rather small. Together, the two brands RCI and Celebrity Cruises attracted only 18 percent of their revenues from non-U.S. sources

(Royal Caribbean, 2007, p. 2). Most of the estimated U.S.$ 940 million came from European customers.

However, from 2006 to 2007, Royal Caribbean Cruises' fleet capacity sailing in Europe has increased from 14 percent to 21 percent, largely due to the addition of Pullmantur. And like Carnival Corporation, Royal Caribbean Cruises have announced their intention to continue increasing their presence in Europe. (RCL, 2007a, p. 4). The planned creation of Croisières de France, a cruise line specifically for the French market (RCL, 2007c), as well as the deployment of seven out of twenty ships of the RCI fleet to Europe in 2008 (RCL, 2007b) underline this commitment.

1.2   Brand, Pricing and Capacity Strategies

Based on the company profiles presented in the previous section we will now proceed to identify the basic patterns behind Carnival Corporation and Royal Caribbean Cruises' observable brand, capacity and price strategies in Europe. Our aim is to reach a level of abstraction suitable for a simple formal analysis of these.

It will prove convenient to choose 2006 as reference year. The reason is not primarily the availability of data for a complete financial or calendar year. More important is the strategy change which Royal Caribbean Cruises seem to have initiated in 2006 with the acquisition of Pullmantur. In our subsequent analysis we will be looking for possible reasons of this change. Therefore we need to focus on Royal Caribbean's pre-change strategy, and the most recent year in which it could be observed was 2006.

Figure 1 gives an overview of the two cruise companies' brand portfolios in the referenceyear . It shows which regional market was served by which brands.

Carnival Corporation

As can be seen in Figure 1, the Carnival Corporation portfolio is composed of brands serving specific regional markets. Only Cunard Line and Costa Cruises transcend the borders of individual markets to a significant degree. Given our interest in the cruise companies' European strategies it is worth noting that historically and culturally, both Cunard and Costa are European brands with traditions reaching back more than 150 years.

Most of Carnival Corporation's twelve brands have preserved a relative independence within the group. "Carnival's unprecedented rise to the world's largest cruise operator can be attributed to its ability to manage brand autonomy, with each major cruise line maintaining separate sales, marketing and reservation offices […]" (Carnival, 2007b). This decentralized structure is characteristic for Carnival Corporation and represents a strategic key element. For our analysis below it is important to highlight that brand autonomy includes pricing autonomy, i.e. each

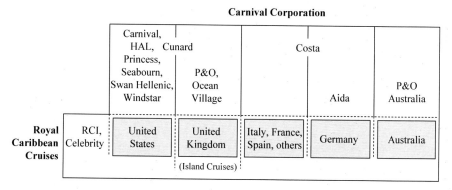

Fig. 1. Segmentation of brand portfolios by market

brand is responsible for all of its own prices and sets them according to its respective strategic objectives and market conditions. Insofar, Carnival Corporation's brand strategy can be called multi-local.

Since cruise ships sail under specific brands, and Carnival Corporation's brands are dedicated to specific markets, it follows that all of the group's ships are also allocated to specific markets. Although transfers of ships from one brand to another, and hence from one market to another, do occur (e.g. the Crown Princess became A-Rosa Blu, then Aida Blu, and now sails as Ocean Village 2), such moves are not very common. At least in the short run, berth capacities are fixed for each brand.

Royal Caribbean Cruises

In contrast to Carnival Corporation, Royal Caribbean Cruises operated only two brands in the reference year, RCI and Celebrity Cruises. Both brands are of U.S. origin but are also marketed outside North America as shown in Figure 1. Four marketing and sales organizations have been set up in the UK and in continental Europe (RCCL, 2007). In 2006, RCI and Celebrity jointly reached a 16 percent market share in the UK and Ireland (Hotel Newswire, 2007). On the continent, however, their market shares were in the lower single-digit range.

Like for Carnival Corporation, increasing their European business is also a priority for Royal Caribbean Cruises. Chairman and CEO Richard Fain stresses that *"We continue to increase our presence in Europe, a key focus of the company."* (RCL, 2007a, p. 4) However, with 82 percent of Royal Caribbean Cruises' revenues coming from the U.S. market in the reference year, their claim to be a global cruise company (RCL, 2007a, p. 27) reflects ambition rather than fact, irrespective of the 50:50 joint venture cruise line Island Cruises in the UK which only represents five percent of Royal Caribbean Cruises' total capacity.

Strictly speaking, the two brands RCI and Celebrity Cruises are not global either. According to Interbrand's definition, a global brand must have a substantial presence in at least one North American, one Latin American, one European and one Asian-Pacific country; at least one third of its business must be outside the home country (for U.S. brands the share may be lower); and a global brand must be managed consistently as such (Interbrand, 2007, p. 53). For the purposes of our analysis, the last criterion is the most important one, and since both RCI and Celebrity Cruises fulfil it, we will refer to these brands as global.

An interesting aspect of Royal Caribbean Cruises' strategy of global brands is the corresponding strategy of global pricing. Essentially, the prices of all products offered worldwide under both brands are set in the corporate headquarters in Miami. Moreover, a particular cabin on a particular cruise booked at a particular moment in time costs the same, wherever in the world the booking may occur. Given this pricing strategy, it is only consequent that Royal Caribbean Cruises hardly ever reserve capacities for individual markets, and that every cabin on every ship can be booked from every market.

1.3   Identifying the High-Level Strategies

At first sight it may seem that the brand, pricing and capacity strategies of the two cruise companies could not be any more different. Carnival Corporation's portfolio of twelve brands, with prices set locally in the respective markets, and capacities inflexibly allocated to defined markets contrast sharply with Royal Caribbean Cruises' two global brands, their centralized, uniform pricing and their globally marketed capacities.

But we may also come to another conclusion. Table 1 shows that the differences listed above can be traced back to only a single difference which lies in the companies' brand strategies, or rather in the brands' market scopes, whereas pricing and capacity allocation follow the same principles at Carnival Corporation and Royal Caribbean Cruises.

As we will see, this single difference is of central importance for the performance of the two companies' strategies in Europe with its many markets varying with respect to consumers' holiday preferences, cultural backgrounds, socioeconomic conditions and competition in the cruise sector.

Table 1. Brand, capacity and price strategies compared

	Carnival Corporation	Royal Caribbean Cruises
Brands	Several local brands. Each brand is offered only in a specific market.	Few global brands. Each brand is offered in all markets.
Capacity	Ships belong to brands and cannot be switched in the short term.	Ships belong to brands and cannot be switched in the short term.
Prices	Set uniformly for each brand.	Set uniformly for each brand.

2   The Model

Based on Table 1, we will now develop a model in order to examine how the two cruise companies' strategies can be expected to perform under European cruise market conditions. As a consequence of the relatively high degree of abstraction from real-world details, and also due to a number of simplifying assumptions yet to be made, the model will necessarily paint a black-and-white picture of the respective strategies' performance and neglect many shades of grey. In order to constantly remind ourselves that the model and the actual companies are not the same, we will refer to the model versions of Carnival Corporation and Royal Caribbean Cruises as "Carnim" and "Royam".

2.1   Assumptions

Our model will be based on five fundamental assumptions about the two companies and the $n>1$ markets in which they operate.

- *Both companies act under identical conditions.* They dispose of the same capacity; compete in the same markets; and face the same price responses from consumers. This assumption allows comparing the strategies on a like-for-like basis since all differences in company performance must necessarily be the result of different strategies and not of different circumstances.

- *Price-response relationships are linear*, which is a common simplifying assumption.

- *The cruise business is a pure fixed-cost business.* The assumption of zero variable costs can be justified on the basis of the relative insignificance of cruise lines' variable costs of sales and food relative to the fixed costs of ship financing and depreciation, payroll, fuel, marketing and administrative overhead. Note that the assumption of zero variable costs implies that maximizing profit or contribution is equivalent to maximizing revenue. In our model, the cruise companies will therefore be revenue maximizers.

- *Within each market, consumers are homogeneous.* Consequently Carnim and Royam will not have to take different segments per market into account.

- *Under each brand, only strictly homogeneous products are offered.* Differences in ship hardware, cabin types, service levels, itineraries, even differences in seasons are excluded. Together with assumption 4 this ensures that each cruise company will only need to set a single price per brand.

Carnim

We will now formalize the model. In the *j*-th market, the price-response function of Carnim's local brand is given by

$$q_j = a_j - b_j p_j \tag{1}$$

where q_j is demand and p_j is price. a_j and b_j are positive, finite parameters determining the position and the slope of the price-response curve, respectively. Since a_j represents demand at a price of $p_j = 0$, it will be referred to as demand potential. The price $\tilde{p}_j \equiv a_j / b_j$ is the lowest price for which $q_j = 0$. It is called the satiating price. Carnim's revenue in the *j*-th markets equals

$$r_j = q_j p_j = (a_j - b_j p_j) p_j. \tag{2}$$

According to Table 1, Carnim manages one brand per market and hence a total of *n* brands. Since the company's cruise ship capacities are dedicated to specific brands and cannot be shifted to other brands easily, Carnim faces one capacity restriction in each market:

$$c_j \geq a_j - b_j p_j \tag{3}$$

where each $c_j > 0$ is fixed in the short run. The sum of all market-specific capacities cannot exceed Carnim's total fleet capacity, therefore

$$c \geq \sum c_i. \tag{4}$$

Royam

For Royam the model differs slightly. Without any loss of generality of the results, Royal Caribbean Cruises' two global brands will be bundled and represented by a single synthetic global brand. Moreover, according to Table 1, Royal Caribbean Cruises do not practice price discrimination among customers of the same brand, no matter which market they come from. Therefore for its global brand, Royam only sets a single price *p* for all *n* markets. The price-response function for the *j*-th market thus reads

$$q_j = a_j - b_j p, \tag{5}$$

giving rise to the revenue function

$$r_j = (a_j - b_j p) p. \tag{6}$$

Given the global nature of their brand, Royam's capacity restrictions are also global and not market-specific like Carnim's. All capacities are offered on all markets implying the following restriction:

$$c \geq \sum (a_i - b_i p). \tag{7}$$

with $c > 0$ being fixed in the short run.

Figure 2 visualizes the two model companies' brand portfolios and their respective segmentation, emphasizing how closely they correspond to their real-world counterparts depicted by Figure 1.

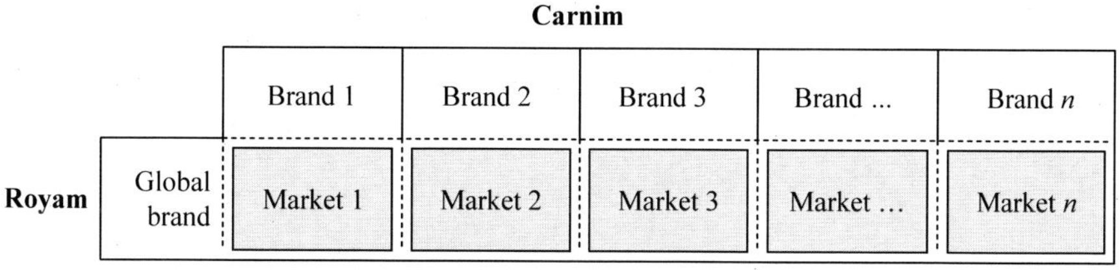

Carnim

		Brand 1	Brand 2	Brand 3	Brand ...	Brand n
Royam	Global brand	Market 1	Market 2	Market 3	Market ...	Market n

Fig. 2. Segmentation of brand portfolios by market in the model

2.2   Brand-Level Optimum

We now have all necessary elements to derive optimal prices, passenger numbers and revenues for each brand. From these results we will be able to draw some first conclusions about the respective strategies' performance in Europe.

Carnim

Carnim's n different brands manage their own prices independently from each other, locally, in a decentralized manner. The j-th brand's objective is to maximize its revenue (2) with respect to its price p_j, subject to its given capacity constraint (3). The maximization procedure is carried out in Appendix A1 and shows that two kinds of optima may exist, an unconstrained and a constrained optimum. In the unconstrained case the brand is endowed with so much capacity that the capacity restriction (3) is irrelevant for pricing. The revenue-maximizing price for the j-th market then equals

$$p_j^* = \frac{a_j}{2b_j}. \tag{8a}$$

The asterisk denotes an optimum. In the second case, demand would exceed capacity if cruises were priced according to (8a). Thus the capacity constraint (3) holds as a strict equality, implying a revenue-maximizing price of

$$p_j^*(c_j) = \frac{a_j - c_j}{b_j}. \tag{8b}$$

which depends on the capacity constraint as indicated by c_j in parenthesis. The capacity restriction becomes effective when $c_j < 0.5 a_j / b_j$; in this case $p_j^*(c_j)$ turns into a linearly decreasing function of c_j. Associated with the unconstrained and constrained optima at brand level are the total passenger numbers and revenues

$$q^* = 0.5\sum a_i \qquad\qquad q^*(c) = c \qquad\qquad (9a/b)$$

$$r^* = \frac{1}{4}\sum \frac{a_i^2}{b_i} \qquad\qquad r^*(c) = \sum \frac{c_i(a_i - c_i)}{b_i}. \qquad\qquad (10a/b)$$

where c is the vector of all brand-specific or market-specific capacity restrictions c_j.

Royam

Let us now turn to Royam. Their single brand's objective consists in maximizing total revenue (6) with respect to the global price p, subject to the capacity constraint (7). Like for Carnim, we assume that all n markets are served by Royam. From the maximization procedure in Appendix A2 follow again an unconstrained and a constrained optimum for price, the number of passengers and revenue:

$$p^* = \frac{\sum a_i}{2\sum b_i} \qquad\qquad p^*(c) = \frac{\sum a_i - c}{\sum b_i} \qquad\qquad (11a/b)$$

$$q^* = 0.5\sum a_i \qquad\qquad q(c) = c. \qquad\qquad (12a/b)$$

$$r^* = \frac{1}{4}\frac{\left(\sum a_i\right)^2}{\sum b_i} \qquad\qquad r^*(c) = \frac{c\left(\sum a_i - c\right)}{\sum b_i} \qquad\qquad (13a/b)$$

Discussion

We have derived Carnim's and Royam's revenue-maximizing prices and passenger numbers, as well as their respective maximum revenues, in Carnim's case taking the allocation of cruise ship capacities to brands as given. The comparison of the results provides a first insight into the implications of the two companies' brand strategies. We can see that *Carnim and Royam transport the same total number of passengers* to maximize their revenues (see (9a/b) and (12a/b)), irrespective of whether or not capacity restrictions apply, as long as the total restrictions are identical for both companies.

We also see that the prices Carnim and Royam charge to achieve these identical passenger numbers, and the revenues they generate are almost certain to differ. Since all results of our model are free from distorting "real-world influences" unrelated to the companies' actions, we can conclude the different prices and revenues can only be attributed to the different strategies. The equations (8a/b) and

(11a/b) show similarities between Carnim and Royam's optimal prices, but more importantly they also show the differences. Carnim's optimal prices are exclusively based on parameters which are specific to the respective local markets: they are fully adapted to local conditions. Royam's global price, on the other hand, takes all n markets' parameters into account and averages them. It is thus a compromise between the specifics of all markets the company serves.

Which company's strategy is superior by leading to a higher total revenue under identical market conditions and with the same available capacities? Let us first look at the unconstrained revenue maxima (10a) and (13a). Appendix A4 proves that (10a) will essentially always exceed (13a). Only in the unlikely event of identical satiating prices $\tilde{p}_j$ in all markets both strategies will achieve the same results. In other words, *if there is abundant capacity for each brand, and satiating prices differ in at least two markets, there is no way Royam can generate as much revenue as Carnim. This superiority of Carnim's strategy becomes even more pronounced the more heterogeneous the n markets are with respect to price response.*

The reason for the superiority of Carnim's brand strategy in the unconstrained case lies in the better adjustability of prices to the respective local market conditions. Royam's global price p^* is too high for some markets and too low for others to be optimal. In both cases Royam foregoes revenue. In view of the heterogeneous European tourism demand this result is quite significant. It suggests that Royam's strategy does not provide the requisite pricing variety to compensate the variety of price responses in the different European markets. This neglect of Ashby's law of requisite variety (Ashby, 1956) may have been without major consequences as long as Royam confined itself to its domestic market and similar markets such as the UK; however, the attempt to penetrate markets worldwide including all major European markets using a global brand strategy and a unique price for all these markets puts Royam in a disadvantaged position compared to its competitor Carnim.

Does Carnim's strategic superiority also carry over to the constrained capacity case? A comparison of (10b) and (13b) cannot give us the answer unless we specify the available capacities c_j for each of Carnim's brands, which we have not done yet. When capacities are scarce, allocating a certain cruise ship to one brand causes opportunity costs for all other brands which could have marketed this vessel, too. Hence economically, the effect of allocating capacity to one of several brands is equivalent to shifting this capacity from one brand to another. The impact of such a capacity shift on Carnim's revenue can be computed using (10b):

$$\left. \frac{dr^*(c)}{dc_j} \right|_{c=const} = \frac{a_j - 2c_j}{b_j} - \frac{a_k - 2c_k}{b_k} \tag{14}$$

where j is the brand receiving the capacity, and k is the brand giving it up. The impact may be positive, zero or negative, depending on the relative magnitudes of all parameters involved. So without information about the concrete distribution of the scarce capacities c_j to brands and markets we cannot make a judgement about the superiority of one strategy over the other.

2.3   Company-Level Optimum

As we have seen from Carnim, optimizing prices for each brand individually leads to local revenue maxima but not necessarily to a total revenue maximum at the level of the entire company. If capacities are allocated inefficiently within the Carnim group the sum of locally maximized revenues may fall short of Royam's total revenue. Hence we will now examine how Carnim ought to allocate its capacities ideally.

For Royam, this question does not arise because there is only one global brand which covers all markets. The company's cruise ships are not reserved for any particular market. Ultimately, the price mechanism decides which ship is filled with which mix of passengers from which markets.

Carnim

Carnim's objective is to find the allocation of capacity c_j which maximizes (13b), given the restriction (4) on Carnim's total capacity. In line with earlier sections, we assume that it is optimal to allocate some capacity to all n markets, i.e. $c_j > 0$. This assumption is equivalent to defining n as the number of markets served by Carnim. Hence we seek to

$$\text{maximize} \quad r^*(c) = \sum \frac{c_i(a_i - c_i)}{b_i}$$

$$\text{subject to} \quad c = \sum c_i .$$

From the maximization procedure in Appendix A3 result expressions for optimally allocated capacities, globally optimal prices and the maximized revenues, all of which now only depend on external parameters of the model:

$$c_j^* = \frac{2b_j c - b_j \sum a_i + a_j \sum b_i}{2 \sum b_i} . \tag{15}$$

$$p_j^*(c_j^*) = \frac{a_j}{2b_j} + \frac{\sum a_i - 2c}{2 \sum b_i} \tag{16}$$

$$r^*(c^*) = \frac{c(\sum a_i - c)}{\sum b_i} + \frac{\sum a_i^2 b_i^{-1} \sum b_i - (\sum a_i)^2}{4 \sum b_i} \tag{17}$$

Obviously, when capacity is scarce it will be fully utilized, so that the optimum numbers of passengers in each market and in total are equal to the respective capacity limits c_j^* and c.

Discussion

Carnim's global revenue maximum requires the company's cruise ship capacity first to be allocated optimally to the n different brands and markets before the prices can be set. Without this two-stage decision making process no globally optimal result can be achieved by Carnim. However, this process is well worth the additional effort because it leads to an outcome that is – again – superior to the best possible outcome that Royam's strategy can produce if in at least two markets the satiating prices $\tilde{p}_j$ differ. Appendix A4 proves that *Carnim's total revenue will not only exceed Royam's if capacities are abundant* (see section 2.2.3) *but also if capacities are scarce and optimally allocated*. As before, the more heterogeneous the n different markets are with respect to their price-response relationships, the better Carnim's strategy will perform relative to Royam's. Only if all n markets are equal, both companies' strategies will achieve the same result.

2.4   Market Specificity of Cruise Products

The demand for cruises reflects specific consumer preferences with respect to brand and product attributes. These preferences may concern languages spoken on board, the nationality mix of passengers and crew, ship hardware, design and decoration, food, formality, entertainment programmes, shore side activities etc. Studies have shown that tourist preferences may strongly correlate with their countries of origin (e.g. TUI, 2003).

In order to have an attractive product to offer, cruise companies adapt their products to their target customers' preferences. For the evaluation of European cruise strategies, the issue of product adaptation is central, given the large number of relatively heterogeneous European markets. Therefore, in this section we will analyze to what extent Carnim and Royam find an optimal adjust of their products to local preferences, given their respective strategies. We will further look at the consequences of product adaptation on prices and revenues. For this reason we have to add a few elements to our model.

We assume that for the j-th market specific consumer preferences can be described by a demand specificity index s_j where $0 \leq s_j \leq 1$. We further assume that cruise companies can shape their products in any way they see fit, and that the product specificity index $\bar{s}_j$ with $0 \leq \bar{s}_j \leq 1$ uniquely expresses a cruise product's profile. Then $(s_j - \bar{s}_j)^2$ measures how well a product profile is adapted to the consumer preferences in the j-th market. In case of full adaptation, $(s_j - \bar{s}_j)^2 = 0$. Finally we suppose that consumers do not only respond to price but also product adaptation: the higher the degree of adaptation to their own preferences, the higher is the demand potential and the satiation price. In our model this effect can be achieved by multiplying the constant parameter a_j of the price-response function by the term $1-(s_j-\bar{s}_j)^2$. In the case of perfect product adaptation $(s_j=\bar{s}_j)$, the price-adaptation-response curve will coincide with the price-response curve of previous sections. Incomplete adaptation, however, will induce a parallel downward shift of the price-response curve or, in other words, a reduced potential demand as well as a reduced satiating price.

Carnim

In market j, Carnim now faces consumer price-adaptation-response which is described by the function $q_j=a_j(1-(s_j-\bar{s}_j)^2)-b_jp_j$. According to assumption 5 (section 2.1), only homogeneous products are offered under each brand. All Carnim brands can thus determine their individually optimal level of product adaptation. In addition to price p_j, the product specificity index $\bar{s}_j$ is now the second decision variable of the j-th brand. Its objective is to

$$\text{maximize } r_j = (a_j(1-(s_j-\bar{s}_j)^2)-b_jp_j)p_j$$

$$\text{subject to } 0\le\bar{s}_j\le1.$$

The capacity restriction (3) has been omitted because nothing new would be gained by re-examining constrained optima. The first order conditions $\partial r_j/\partial p_j=0$ and $\partial r_j/\partial\bar{s}_j=0$ which are also sufficient for a unique global maximum can be written as

$$p_j^*(\bar{s}_j)=\frac{a_j(1-(s_j-\bar{s}_j)^2)}{2b_j} \tag{18}$$

$$s_j^*=\bar{s}_j \tag{19}$$

Condition (19) says that it is optimal for each Carnim brand to fully adapt its products to consumer preferences of its respective market. Upon substitution from (19), condition (18) reduces to (8a). Also optimal passenger numbers and revenues turn out to be equal to their basic counterparts (9a) and (10a). It follows that Carnim's optimum is not affected by the introduction of product adaptation possibilities into the model.

Royam

Due to Royam's global brand, there is not only one global price p but also only one globally homogeneous product profile (again see assumption 5 in section 2.1) denoted by $\bar{s}$. From Royam's unconstrained optimization problem to

$$\text{maximize } r = p\sum(a_j(1-(s_j-\bar{s})^2)-b_jp)$$

$$\text{subject to } 0\le\bar{s}\le1.$$

follow the necessary and sufficient conditions $\partial r / \partial p = 0$ and $\partial r / \partial \bar{s} = 0$ which can be rearranged to read

$$p^* = \frac{\sum a_i - \sum a_i (s_i - \bar{s})^2}{2 \sum b_i} \qquad (20)$$

$$\bar{s}^* = \frac{\sum a_i s_i}{\sum a_i} \qquad (21)$$

Condition (21) specifies the optimal $\bar{s}$ as the mean value of all s_j, weighted by the relative potential demand $a_j / \sum a_i$ in each market respectively. This means that Royam gives the demand profiles of markets with a larger demand potential more consideration in its product adaptation decision than the demand profiles of smaller markets. If we define the variance $\sigma^2 \equiv \sum a_i (s_i - \bar{s})^2 / \sum a_i$, we can re-state (20) as

$$p^* = (1 - \sigma^2) \frac{\sum a_i}{2 \sum b_i}. \qquad (22)$$

This says that incomplete product adaptation drives Royam's global price down below its (11a) level. But while the price is decreasing only linearly in $(1-\sigma^2)$, revenue is decreasing even quadratically:

$$r^* = \frac{1}{4} \frac{((1-\sigma^2) \sum a_i)^2}{\sum b_i}. \qquad (23)$$

Discussion

This short analysis has shown that market-specific brands, products and prices are ideally suited to deal with a diversity of consumer preferences such as we encounter in Europe. Carnim's multi-local brand strategy provides the requisite variety to counter the demand variety of heterogeneous markets. This explains its superiority over Royam's global brand strategy which performs best across markets with similar consumer preferences.

Moreover, if the markets in which Royam operates differ significantly with respect to their demand potential, Royam's product profile will be very strongly influenced by the demand profile of those markets or of the single market with the largest demand potential. Given the extraordinary size and the demand potential of the U.S. cruise market, our model suggests that *Royam's product profile will be almost perfectly adapted to the U.S. passengers' preferences whilst hardly reflecting smaller markets' needs*, a result which seems to correspond very closely to the actual situation.

This last point needs to be stressed a little more as it has very important strategic consequences. If Royam's global brand and its products are indeed less attractive for many Europeans, and Royam wants to change this, the company will find itself in a dilemma. *By making its product profile "less American", whatever this may mean, Royam can win customers in Europe while losing customers in the U.S.* But why should the company accept this trade-off?

Our model suggests that trading off two large demand potentials against one another is unnecessary. As the consistent superiority of Carnim's multi-brand strategy has demonstrated *there is a better way for Royam to compete in Europe: by limiting the presence of its brand to clusters of markets with similar preferences and by pursuing a multi-local brand strategy for the remaining markets.*

3   Back to Reality

In this final section we will look at further implications of our model, point out why some of its outcomes should be taken with a grain of salt, and attempt an interpretation of Carnival Corporations' and Royal Caribbean Cruises' recent strategic moves in the light of our results which are summarized in Table 2.

Table 2. Model scenarios and best-performing strategies

Revenue maximization	Superior performance in the …	
	unconstrained optimum	constrained optimum
… at brand level	Carnim's strategy	Undetermined
… at company level	Carnim's strategy	Carnim's strategy
… with product adaptation	Carnim's strategy	Not modelled

The superiority of Carnim's brand strategy is not limited to the scenarios examined with our model. Carnim's strategy can also be shown to master various related situations better than Royam's strategy. Three examples can be given here:

- *Dynamic pricing*: Royam's global price can lead to difficulties in the short-term fine-tuning of prices during the booking period. Since price-response relationships change as the departure date of a cruise approaches, and these changes are likely to differ across markets (e.g. some markets tend to book earlier than others), dynamic pricing with global prices is problematic.

- *Different local competitive situations*: The same company may hold a solid market-leading position in one market and be a small newcomer in another. In the first case, skim pricing might be the best strategy, while the second case may require penetration pricing. Royam's global price cannot cope with both situations simultaneously in a satisfying way, nor does it allow responding appropriately to local competitors' moves.

- *Different objectives for different markets*: Even in largely homogeneous markets, Royam's global brand strategy may represent a handicap for the company because it hardly permits the pursuit of different objectives in different markets. Prioritizing one market over another requires treating them differently which is difficult given a global brand, global products and globally uniform prices.

After pointing out the most serious shortcomings of Royam's global brand strategy relative to Carnim's multi-local strategy within our analytical framework, we need to remember that the model and reality are not the same, and that our results need to be put into perspective. For instance, unlike Royam, the real-world company Royal Caribbean Cruises can decide to protect some of its capacity for a priority market and manage the prices for this allotment separately, thus mitigating some disadvantages of globally uniform pricing. Also unlike Royam, Royal Caribbean Cruises do adjust their products to the demand profiles of European markets by deploying ships to Europe and by offering a selection of onboard services in languages other than English. And if our model allowed for positive scale effects, which are a major argument in favour of a global branding, the superiority of Carnim's strategy may have turned out less unambiguous than it did.

We should also reconsider Carnim's two-stage decision making process (capacity allocation before price-setting) under more realistic conditions. Clearly, this process is much more demanding, requires much more information, and is much more susceptible to mistakes with longer-term consequences than Royam's approach. The question arises whether at all it is possible for Carnival Corporation to achieve at least a near-optimal allocation of cruise ship capacities in practice, given that customers' responses to prices and products are subject to many external influences, that they tend to change, and they can only be known with a degree of uncertainty. Carnival Corporation's policy of setting hurdle rates of return which brands must earn to qualify for additional capacity is surely a good rule of thumb but it is still based on historical data and extrapolation and can therefore not guarantee optimal forward-looking capacity allocation like in our model.

How can Carnival Corporations' and Royal Caribbean Cruises' recent strategic moves be interpreted in the light of our findings? Despite Royal Caribbean Cruises' ambition and efforts to increase their presence in Europe with their existing brands RCI and Celebrity Cruises, their success in continental Europe has been limited. According to the discussion in section 2.4 this may at least partly be a consequence of inadequate product adaptation and of prices which do not sufficiently reflect specific local competitive circumstances.

It is too early to call the 2006 acquisition of Pullmantur in Spain by Royal Caribbean Cruises a strategic turning point. But if Pullmantur is indeed kept as an independent brand within Royal Caribbean Cruises' portfolio, as Acher and Reiter (2007) report, the acquisition marks the beginning of the end of the company's global brand strategy. Royal Caribbean Cruises' announcement to create a new cruise brand exclusively for the French market seems to confirm this strategy change which looks like an attempt to overcome the problems of global brands we also encountered in our model.

With dedicated brands in European key markets – not only France and Spain – with their joint venture Island Cruises, their two existing global brands RCI and Celebrity Cruises, and with the deluxe brand Azamara Cruises, Royal Caribbean Cruises could finally match Carnival Corporation with respect to local product adaptation and pricing capabilities. Their new strategy could be called "glocal" (e.g. Holt *et al*, 2004), a hybrid of global and local.

Interestingly, Carnival Corporation appears to be converging towards the same glocal strategy but from the opposite side. Under the Carnival roof, Costa Cruises have developed from a local Italian brand to a pan-European brand. And while Costa Cruises display their global ambitions by expanding to China, Carnival Corporation co-creates the new local Spanish cruise line Iberocruceros.

The world's leading cruise companies finally seem to agree on a formula for success: as global as possible, as local as necessary. But what is possible? And what is necessary? Europe might be just the perfect testing ground for them to find it out.

References

Acher, J. and Reiter, C., 2007, *UPDATE 4-Royal Caribbean buying Spanish cruise operator*. http://www.reuters.com/article/idUKL3130919320060831?sp=true (accessed 14 Sept. 2007).

Ashby, W. R., 1956, *Introduction to Cybernetics* (London: Chapman & Hall).

Carnival, 2007a, *Carnival Corporation & plc Annual Report 2006*. http://library.corporate-ir.net/library/14/140/140690/items/234461/CCP06AR.pdf (accessed 13 Sept. 2007).

Carnival, 2007b, Mission and History. http://phx.corporate-ir.net/phoenix.zhtml?c=200767&p=irol-history (accessed 13 Sept. 2007).

Carnival, 2007c, *Carnival Corporation & Plc and Orizonia Finalize Agreement for New Spanish Cruise Operation*. Press release of 5 September 2007. http://phx.corporate-ir.net/phoenix.zhtml?c=200767&p=irol-newsU.S.CCL&nyo=0 (accessed 14 Sept. 2007).

CLIA, 2007, *Study results detail cruise industry's $35.7 billion dollar contribution to U.S. economy*. Press release of 29 Aug. 2007 by the Cruise Lines International Association. http://www.cruising.org/cruisenews/news.cfm?NID=273 (accessed 26 Sept. 2007).

Ebersold, B., 2007, Europe fuels industry optimism. *Marine Log*, June, 31-34 http://www.nxtbook.com/nxtbooks/sb/ml0607/index.php (accessed 26 Sept. 2007).

Holt, D. B., Quelch, J. A., Taylor, E. L., 2004, How Global Brands Compete. *Harvard Business Review*, 82 (9), 68-75.

Hotel Newswire, 2007, *Royal Caribbean Restructures in UK*. http://www.hotelexecutive.com/newswire/pub/_24635.asp (accessed 14 Sept. 2007).

Interbrand, 2007, *All Brands Are Not Created Equal. Best Global Brands 2007*. http://www.ourfishbowl.com/images/surveys/Interbrand_BGB_2007.pdf (accessed 9 Sept. 2007).

Mäkinen, E., 2007, *Shipbuilding and Cruise Industry*. Document presented on 6 February 2007 at the European Cruise Conference in Brussels. http://www.europeancruisecouncil.com/eci/ECIConferenceReport/Shipbuilding%20seminar/ (accessed 21 Sept. 2007).

Mathisen, A. R. and Mathisen, O., 2007, Market Leader. *Cruise Industry News Quarterly*, summer, 86-88.

RCCL, 2007, Über uns. http://www.rccl.de/ueber_uns.htm (accessed 14 Sept. 2007).

RCL, 2007a, *Royal Caribbean Cruises Ltd. Annual Report 2006*. http://library.corporate-ir.net/library/10/103/103045/items/248875/06AnnualRep.pdf (accessed 13 Sept. 2007).

RCL, 2007b, *Royal Caribbean Deploys Record Seven Ships To Europe In Summer 2008*. Press release of 9 March 2007. http://www.royalcaribbean.com/ourCompany/pressCenter/pressReleases.do (accessed 14 Sept. 2007).

RCL, 2007c, *Royal Caribbean Cruises Ltd. Starts New Cruise Line Dedicated To French Market*. Press release of 13 Sept. 2007. http://www.royalcaribbean.com/ourCompany/pressCenter/pressReleases.do (accessed 14 Sept. 2007).

Sbarsky, A., 2007, Europe: Cruising's Hot Spot. *Cruise Industry News Quarterly*, summer, 72-77.

TUI, 2003, *Destination Images*. Unpublished study commissioned by TUI AG, Hannover.

Appendix

Appendix A1

From the Lagrangean function $L_j = (a_j - b_j p_j)p_j + \lambda(c_j - a_j + b_j p_j)$ follow the necessary conditions for a unique global maximum which are also sufficient:

$$\frac{\partial L_j}{\partial p_j} = a_j - 2b_j p_j + \lambda b_j = 0 \tag{A1}$$

$$\frac{\partial L_j}{\partial \lambda} = c_j - a_j + b_j p_j \geq 0, \ \lambda \geq 0 \ \text{and} \ \lambda \frac{\partial L_j}{\partial \lambda} = 0.$$

If $c_j > a_j - b_j p_j$ then the Lagrangean multiplier λ vanishes and (A1) can be solved for

$$p_j^* = \frac{a_j}{2b_j}. \tag{A2}$$

On the other hand, if $\lambda > 0$ then $c_j - a_j + b_j p_j = 0$. Solving it for the price leads to

$$p_j^*(c_j) = \frac{a_j - c_j}{b_j}. \tag{A3}$$

Substituting from (A2) and (A3) into (1) and summing over all n markets leads to (9a) and (9b). Substituting from (A2) and (A3) into (2) yields (10a) and (10b).

Appendix A2

From the Lagrangean function $L = p\sum(a_i - b_i p) + \lambda(c - \sum(a_i - b_i p))$ follow necessary and sufficient conditions for a unique global maximum:

$$\frac{\partial L}{\partial p} = \sum(a_i - 2b_i p) + \lambda \sum b_i = 0$$

$$\frac{\partial L}{\partial \lambda} = c - \sum(a_i - b_i p) \geq 0, \; \lambda \geq 0, \text{ and } \lambda \frac{\partial L}{\partial \lambda} = 0$$

Reworking the same steps as in Appendix A1 leads to (11a/b), (12a/b) and (13a/b).

Appendix A3

The Lagrangean function reads $L = \sum c_i(a_i - c_i)/b_i + \lambda(c - \sum c_i)$. Its maximization yields the necessary and sufficient condition for a unique global maximum:

$$\frac{\partial L}{\partial c_j} = \frac{a_j - 2c_j}{b_j} - \lambda = 0$$

Equating two such conditions for different markets j and k shows that

$$c_k = \frac{a_k b_j - (a_j - 2c_j)b_k}{2b_j}.$$

Substituting the result into the capacity constraint (4) and rearranging terms yields (15) which upon substitution into (11b) and (13b) leads to (16) and (17).

Appendix A4

Carnim and Royam's unconstrained revenue maxima are given by (10a) and (13a), respectively. Deducting the latter from the former yields

$$\frac{1}{4}\sum\frac{a_i^2}{b_i} - \frac{1}{4}\frac{\left(\sum a_i\right)^2}{\sum b_i} = \frac{\sum a_i^2 b_i^{-1} \sum b_i - \left(\sum a_i\right)^2}{4\sum b_i}. \tag{A4}$$

If Carnim's unconstrained revenue maximum is superior to Royam's, this expression must always be non-negative and at least in one case positive. Interestingly, Carnim and Royam's constrained revenue maxima given by (10b) and (13b) differ only by one additive term, and this term is identical with (A4). Therefore, to prove the general superiority of Carnim's revenue maximum it suffices to show that (A4) is non-negative and in at least one case positive. To this end, we rearrange the term as follows:

$$\frac{\sum a_i^2 b_i^{-1} \sum b_i - (\sum a_i)^2}{4 \sum b_i} \geq 0 \Leftrightarrow \sum \frac{a_i^2}{b_i} \sum b_i - (\sum a_i)^2 \geq 0$$

$$\Leftrightarrow \sum \frac{a_i^2 \sum b_j}{b_i} - (\sum a_i)^2 \geq 0 \Leftrightarrow \sum a_i^2 + \sum \frac{a_i^2 \sum_{j \neq i} b_j}{b_i} - \sum_i (a_i \sum_j a_j) \geq 0$$

$$\Leftrightarrow \sum a_i^2 + \sum \frac{a_i^2 \sum_{j \neq i} b_j}{b_i} - \sum a_i^2 - \sum_i (a_i \sum_{j \neq i} a_j) \geq 0$$

$$\Leftrightarrow \sum \frac{a_i^2 \sum_{j \neq i} b_j}{b_i} - \sum_i (a_i \sum_{j \neq i} a_j) \geq 0 \Leftrightarrow \sum \frac{a_i^2 \prod_{j \neq i} b_j \sum_{k \neq i} b_k}{\prod_j b_j} - \sum_i (a_i \sum_{j \neq i} a_j) \geq 0$$

$$\Leftrightarrow \sum_i (a_i^2 \prod_{j \neq i} b_j \sum_{k \neq i} b_k) - \prod_j b_j \sum_i (a_i \sum_{k \neq i} a_j) \geq 0$$

$$\Leftrightarrow \sum_i \sum_{j \neq i} (a_i b_j)^2 \prod_{k \neq i, j} b_k - \prod_k b_k \sum_i \sum_{j \neq i} a_i a_j \geq 0$$

$$\Leftrightarrow \sum_i \sum_{j \neq i} (a_i b_j - a_j b_i)^2 \prod_{k \neq i, j} b_k \geq 0$$

Since $(a_i b_j - a_j b_i)^2$ cannot be negative and $b_j > 0$ by assumption, the whole term is non-negative. Moreover, if at least one parametric combination exists for which $a_i / b_i \neq a_j / b_j$, $\Leftrightarrow \tilde{p}_i \neq \tilde{p}_j$ then $(a_i b_j - a_j b_i)^2$ is positive.   q.e.d.

Business Travel Management

Current Developments in the Business Travel Sector

Andreas Wilbers

1 Introduction

Economies have been flourishing in recent years leading to a positive development of the business travel sector. Business travel, particularly to the booming regions of Asia (especially China and India) and Latin America (for example Brazil), poses new challenges to travel managers. Whereas the focus of international travel management used to be on the old industrial nations of Europe and North America, the impact of Asian countries has increased considerably over the past few years. However, Arab countries also want to benefit from the economic boom and have been busily engaged in the building of new tourism capitals with increasingly extravagant and modern hotels and airports. The periods of economic growth were abruptly interrupted by the financial crisis in the middle of 2008 and it remains to be seen which, if any, of the service providers actively involved in the business travel sector can cope with the new challenges. On the other hand, travel managers are once more gaining in importance in their capacity as buyers. Excess supply changes the balance of power between service providers and travel managers in favour of the latter. The following report describes current developments in the business travel sector as well as options available to the actors involved in it.

2 Developments in the Airline Industry

2.1 Power Struggle Between Airlines and GDS

In Europe, a serious power struggle has broken out between airlines and Global Distribution Systems (GDS). Airlines in the United States had already ensured that it would only be possible to book their cheapest fares via GDS if the GDS fees were decreased considerably. The particularity of the European market is that Lufthansa, Air France and Iberia still hold shares in Amadeus GDS. Lufthansa also tried through intensive negotiations with Amadeus in 2008 to achieve lower GDS booking fees. Amadeus, however, absolutely rejected this. Since Lufthansa, as Amadeus shareholder, could not simply remove the cheapest fares from the GDS, the airline took the roundabout way by offering travel agencies so-called preferential prices. These, however, are only available in Amadeus in exchange for a separate GDS payment. In this way, Lufthansa tried to use the travel agencies to put pressure on Amadeus. But the business travel agencies in Germany are in no position to put pressure on Amadeus as it has a virtual monopoly in the business travel sector. In the end, the GDS fees hit the travel managers in the businesses. An important lesson to be drawn from this situation, which has been difficult for both travel agencies and their clients, is that, along with Amadeus, travel agencies must have further GDS in their portfolio. Travel agencies and their clients can only break the unwanted yet secretly existing alliance between Lufthansa and Amadeus by increasing the competition.

2.2   Airline Surcharges

Airlines are becoming ever more creative at inventing additional charges for flights. It is becoming increasingly difficult for business travellers to calculate the total cost of a flight. For example, luggage allowances vary from one airline to another. Furthermore, some airlines want their customers to indicate the amount and the total weight of the checked luggage at the time of booking. Excess luggage booked in advance, incurs smaller fees than luggage checked in at the time of flying. Additional charges for seat reservations and credit card fees are a further source of irritation for many travellers. From a business point of view, it is difficult to comprehend why airlines should wish to come up with inefficient processes that only succeed in annoying business travellers. It is also barely comprehensible why a business traveller with a mere 2kg of excess luggage at check in should be expected to go to another desk, pay a few euros and then have to return to the check-in desk to start queuing all over again. It is not until this rigmarole has been completed that he can receive his boarding card. Credit card surcharges are just as absurd. Why is it that an airline can demand remuneration for a means of payment which to them is almost risk-free while the less secure payments made by direct debit are free of any surcharges. The list of surcharges is endless and it may be only a question of time until using the plane toilet will be used as an excuse for extracting a pee-fee.

3   Corporate Social Responsibility

3.1   Alternatives to Business Travel

Business trips entail high costs. More and more companies are looking for ways to reduce the business expenses entailed in communication. Telephone, video and Internet conferences are accepted alternatives to business trips. But, how can business travellers actually be motivated to use these alternatives effectively? Self-booking tools (SBT) offer an interesting means of assistance. If a traveller wishes to book a flight, the SBT initially asks if other alternatives to the trip are available. A simple "yes" or "no" will not suffice, however, and an additional reason must be given briefly. Only after this field has been completed is it possible to proceed to the flight reservation mask. This ensures that travellers are forced to consider the necessity of their journey. After all, the boss can access the forms at any time.

But in many cases it is not enough to just point out alternatives to business trips. Training on how to professionally organise telephone, video and Internet conferences is something that is completely neglected. Most users are left alone and already feel stressed by the preparation phase of such conferences. During the conference, the moderator not only has to lead the conference but must also ensure that every participant is given the opportunity to express his opinion. Language problems are often a feature of international conferences and while, in a face-to-face situation, mimic and gesture can often overcome such problems, this is only partially possible in video conferences and does not work at all in the case of telephone or web conferences. Small wonder then that some users at these kinds of conferences simply make their excuses and leave after a short time.

3.2 Environmental Aspects

The global financial crisis has ensured that environmental issues have been neglected throughout the second half of 2008. The enormous bailout packages for the financial markets are threatening the urgently needed investment in environmental protection. Car sales have dropped drastically and therefore, carmakers are investing less money in the development of environmentally friendly cars. Airlines are postponing the purchase of new, fuel-saving aeroplanes. It all amounts to a vicious circle that is only going to be broken when the economy recovers. The alternatives to business travel outlined above can help companies to cut travel expenses, as well as to save the environment, as fewer business trips mean fewer emissions. However, the importance of personal contact between people should not be underestimated: it helps to build confidence, which is an important prerequisite for new business relationships. Avoiding travel should not become an end in itself, however, but should be seen as a useful measure to help cut expenses and protect the environment. With governmental support manufacturers must push forward development towards the goal of environmentally friendly transportation.

3.3 Travel Security

Worldwide terrorism is paralysing the business travel market in many parts of the world.

Security aspects play an important role for airlines in particular. The European Union has been desperately trying to find ways of reducing the inconveniences for business travellers. However, their methods have sometimes been rather questionable. The development of a full body scanner, for example, to enable airport security staff to scan through passengers' clothes is from an ethical point of view a more than dubious intervention. On the other hand, some business travellers find having to undergo a body search disagreeable and object to having to take their shoes and belt off in front of others. Afterwards, they need to redress quickly because the next travellers are already waiting for this procedure. The resulting stress as well as the loss of time caused by early check-in and time-consuming security checks have contributed to increased travelling costs over the last few years.

However, flights are not the only area where security plays an important part. The security situation in developing countries can often not be compared to that in industrial countries. Even a taxi ride may turn out to be life threatening. Politicians from the industrial nations should demand higher security standards in those countries. For example, development aid could be linked to improvements in the security sector.

4   Technology

4.1   The Influence of Self-Booking Tools on Travel Management Companies

Businesses are looking for alternatives to the classic travel agency in order to absorb the increasing costs for such agencies. The most important alternatives are self-booking tools (SBT). Self-booking tools are not yet being widely used. Travellers and secretaries' offices often call the travel agencies or send an email with their reservation requests. However, what is starting up slowly today may well become standard practice tomorrow. After all, more and more businesses are already planning to use these systems in the future.

What can these systems do? They are well suited for the booking of standard trips. Booking hotel rooms and rental cars as well as searching for train connections is no longer a problem. Domestic and simple international return flights as well as the booking of corporate net rates via SBTs are already being offered to travellers as a standard by some businesses. As these travellers are no experts on travelling rates, many bookings have to be rechecked for optimisation potential. Flight tickets can then be issued electronically, customized invoices created and the relevant information entered into the management information system. Therefore, some of the classic tasks of travel agencies will remain. However, one has to assume that these tasks will be automatised more strongly and carried out with considerably less staff than is the case today.

Nowadays, a professional travel agency has average costs of 5% to 7%, depending on the structure of their collaboration with businesses. By using SBTs these costs should be cut in half. In other words, using an online booking system means a travel agency requires only half the staff. However, there are also new tasks for the travel agencies. Businesses, for example, require help with the implementation and regular maintenance of the systems; training for the travellers has to be provided. In short, the travel agencies have to become close partners of the businesses, which are becoming less dependent on the service providers. Those who miss out on this development are going to have to concentrate in the future on the few businesses that do not want to use online booking, either that, or expect, sooner or later, to be closing their own business down.

Now one might ask whether the employees of travel agencies are being sufficiently trained for their new field of activity? This question must be answered with a clear "no". At the moment employees are only trained for the work in a classic travel agency. Further training needs to be provided by the agencies. They need to know how the different self-booking tools work so they can be used to help the businesses to adjust to their individual requirements. In addition, they have to be able to understand the basic problems of their clients, which go beyond just wanting the cheapest fares. The German Business Travel Association (VDR – Verband Deutsches Reisemanagement) reacted to this development several years ago and founded an academy for travel management. Originally focusing on the businesses, this academy has, from the beginning, also been attended by travel agencies and service providers. The seminars for certified travel management enable the travel agency employees to gain a deeper insight into the demands of the businesses. Michael Kirnberger, chairman of the VDR, states: "In the future, the collaboration between travel agencies and businesses will turn out to be much closer and a lot more cooperative. Travel agency employees are already valued guests at our seminars. During the last years the academy has considerably expanded the range of seminars for this target group."

The travel agencies have no time to lose in adjusting to the changing requirements. Within the next few years the market for business travel will change considerably. This means making use of as many opportunities as possible for advanced professional training. Instead of study trips the focus has to be put on advanced training in travel management. Techniques such as project management or business process optimisation as well as an understanding of the possibilities of the new media are skills which will be required for future travel agency employees. New professional titles such as travel administrator and system administrator will find their way into the travel agencies. The work will become more demanding and the salary level for specialists will increase. Travel agencies will not become extinct; however, some of their work will change fundamentally.

4.2   Quality Control Systems (Robotic Software)

Ms Smith has just booked her business flight to New York. Unfortunately, she will have to take an earlier flight, since her preferred flight is already fully booked. However, her travel agency has reserved her a place on the waiting list. Ms Smith prefers to have an aisle seat, but at the time of booking only a middle seat was available. In addition, the fare confirmed by the travel agency was considerably higher than for previous flights.

This typical example shows the difficulties business travellers have to deal with. A professional travel agency will still try everything to satisfy Ms Smith's wishes. A travel agency employee will regularly search the GDS in order to find an available seat for her on her preferred flight. By occasionally checking the system, they might even happen to find one of the favoured aisle seats. But this sort of service requires a lot of time and effort and drastically reduces the travel agent's productivity. Even selling a ticket to New York can become a loss-making transaction in these circumstances. Moreover, the work climate in the agency is hardly going to benefit if the customers' wishes cannot be satisfied at all or only with great effort.

However, for some time now there has been software available that can help to process Ms Smith's order without too much additional work. These computer programs are known as robotic software or quality control systems. They were originally developed for the US market. The majority of flights booked by American business travellers are domestic flights. Although there is a great variety of fares for domestic flights, offers are in essence quite simply structured; American software developers were thus able to create the robotic software to save travel agents a great deal of the time and effort previously spent in looking for the cheapest fare. A few years later, this software was introduced in Europe, too. However, it was not widely used at first, because in countries such as Germany where most tickets are sold for international flights the software had only limited application. Today robotic software is also being developed in Europe and has reached very high standards. All of these programs basically use the same method; they regularly and automatically check whether a customer's request concerning waiting lists, seat preference, and lower fares can be met and go on doing so until shortly before the flight is due to take off. As soon as there is a positive response to the query, the software will notify the travel agent who can then immediately inform the customer.

How exactly do these programs work? They first check that all information necessary for a booking has been provided. To do this, different options can be set. If a company wants to analyse its accounts on a monthly basis, the program will check whether the travel agent has entered the cost centre correctly.

As soon as a waiting list is booked, the program queries the computer reservations system for available seats on a regular basis (for example every hour, every two hours or twice a day). Once a seat becomes available, the waiting list is cleared and the seat booked. The same method is applied to search for preferred seat assignments. It is even possible to leave the seat reservation completely to the system, so that the travel agents do not have to book the seats themselves.

The airlines' fares are constantly changing. It is impossible for the travel agencies to continuously check all the bookings in search of better fares. Here, too, the robotic software helps by searching the computer reservations system for lower fares at pre-assigned intervals. Another advantage is that it is possible to search different GDS (for example Amadeus, Galileo, Sabre, Worldspan) at the same time.

Further developments in robotic software, for example enabling it to search for the lowest fare while taking account of creative ticketing, would considerably facilitate work in company travel departments. The continually rising cost pressure can only be decreased by using complex computer programs. Unfortunately, the potentials of robotic software are not yet being intensively or widely enough exploited. Robotic software is a tool that helps travel agents offer their customers high-quality service. This technology provides the best buy guarantee customers keep demanding.

> And what about Ms Smith? She is happy that her travel agency was able to secure her a seat for the flight she wanted. Shortly before departure, she was even able to change her booking to a lower booking class, which saved her a few hundred Euros. Her aisle seat was also confirmed in advance. In short, without the quality control system, her travel agency would not have been able to meet Ms Smith's booking preferences.

4.3   Data Security in Business Travel

Due to worldwide terrorism, airport security measures have been enhanced not only for passengers and luggage, but also for data storage devices. Because authorities want to prevent the distribution of hazardous data, U.S. Customs and Border Protection (CBP) and other agencies now have the right to seize travellers' laptop computers or to make backup copies of hard drives. This opens the doors to government-organised industrial espionage, a development that is unacceptable for business travellers. Unfortunately, European authorities do not have the power to prevent or alleviate these measures introduced by the USA. The only thing that would help would be the taking of decisive action against these measures. The same applies to Asian countries.

4.4   Management Information Systems (MIS)

The efficient evaluation of business travel data is still an unsolved problem. Different sources such as credit cards, travel agencies, GDS, SBT and service providers deliver extensive data. However, travel managers still have the problem of processing data efficiently. It often gets confusing when one single query yields different evaluation results. For example, it has long been a source of puzzlement to travel managers why it is that the credit card evaluations for air travel provide different results from the evaluations of the travel agencies; even though all bookings were made in the travel agency and with one single credit card.

For hotel evaluations, important data are missing. Credit cards deliver only one figure per accounting record. Therefore, the number of nights the traveller stayed in the hotel remains unknown. Neither are there any details of the different cost types. Travel agencies and hotel portals only deliver booked, but not actually invoiced figures of overnight stays. No-shows do not appear at all. The hotel industry needs to get its act together, so that the clients are finally provided with reliable data about all data sources.

The deficiencies that arise with national MIS are intensified for internationally operating travel managers. Often not even the service providers are able to provide their clients with reliable data. But if this is not possible, how can efficient, global sales and purchase negotiations be managed? Some companies have therefore already decided to abandon their lofty ideals of international consolidation. They await the dawn of better MIS times and in the meantime are limiting their international role to the coordination of worldwide business travel activities, for example by introducing and managing global travel policy guidelines.

5   Demands on Travel Management Companies

As part of a research study, the author of this article investigated the future demands on business travel agencies.

With the help of a model developed by the Japanese researcher Kano service requirements can be divided into three categories:

Table 1. Categories of customer requirements

Basic requirements	*Must be* service characteristics. Their absence results in extreme customer dissatisfaction.
Performance requirements	Customer satisfaction depends on how well these demands are met: The better the performance, the more satisfied the customer will be.
Excitement requirements	Non-fulfilment does not lead to customer dissatisfaction. However, fulfilment of these demands results in high levels of customer satisfaction.

Based on the Kano Model, it was necessary in analysing the future service requirements that will have to be met by business travel agencies, to consider these three different categories of customer needs. However, a further category must also be taken into consideration, namely the indifferent, or non-committal, attitude on the part of the respondents. The service is not expected, but neither is it viewed negatively.

The empirical survey was conducted from November 2007 to January 2008. It represents approximately 21.8 % of the business travel turnover in Germany controlled by travel management. The results of the survey can be summed up as follows:

1. **Result:** When polled, the companies divided their needs into three types of requirements: basic requirements (must be), excitement requirements (attractive), and requirements rated as indifferent. The fact that one-dimensional services (performance requirements) were not assigned, suggests that companies desire neither the performance requirements to be met proportionally nor consequently the performance quality. If a performance attribute is offered, then it should also be of a high quality.

2. **Result:** As basic requirements, the classic services of a travel agency, such as the offer of different ways of booking, fast response to booking requests, courier services, and information on entry requirements were named. Travel reservation is therefore very much the central focus. When booking business travel, companies expect high quality performance. This performance is taken for granted and even where standards are met, it simply results in the companies not being dissatisfied.

3. **Result:** Service features rated as indifferent, for example, advisory services, offers of video conferencing systems, vehicle fleet management, travel insurances, give the travel agencies the opportunity of differentiating themselves from their competitors by compiling customised service packages.

4. **Result:** The assignment of highly qualified staff, that is to say, trained travel agents for bookings (basic requirements), graduates for account management and consulting services (excitement requirements), is desired. In particular, the assignment of higher qualified staff, however, still constitutes a problem for business travel agencies. The classic career in a travel agency starts with an apprenticeship and is followed by working one's way up through the hierarchy.

5. **Result:** A segment-specific evaluation according to company size revealed differing views among respondents on several service features. For instance, obtaining of visas: rated as indifferent by SMEs, but as a basic requirement as far as large-scale enterprises were concerned; determining position of travellers in times of crisis: excitement requirement for SMEs, but rated a basic requirement among large-scale enterprises. It is therefore possible to differentiate between travel agencies on a segment-specific basis. Further segment-specific studies have not been conducted, but would be conceivable, for example, with regard to the international orientation of travel management or industry-specific clusters.

6. **Result:** The research has shown that business travel agencies can establish barriers to market entry by utilizing the full range of services and that effective instruments are at hand to face the dangers resulting from substitute products and the bargaining power of both clients and suppliers as well as the competition with established agencies.

- **Barriers to entry**: e.g. reduction of company size related costs through merging of facilities;

- **Competitive pressure through substitute products**: e.g. on-line booking tools provided by travel agencies;

- **Bargaining power of clients**: e.g. provision of customized service packages;

- **Bargaining power of suppliers**: comprehensive service packages vs. direct sales;

- **Competition between established competitors**: differentiation strategy (e.g. distinctive service packages), cost leadership regarding basic requirements (e.g. booking processes), concentration on core businesses (e.g. geographic markets), time advantages (cutting-edge advisory services).

7. **Result**: Personnel strategic options play an especially important role when it comes to resource-oriented implications. The survey has indicated that the demand for specially qualified travel agency personnel is very high. Travel agencies can only establish and extend their core competencies, if they are ready to employ highly qualified staff. Therefore, the travel agencies must prioritise the employment of graduates as well as the instruction and further training of their staff. A travel agency does not possess any core competence, unless it offers services that, aside from customer benefits, also imply distinctions from the competition and the further development of their core competencies. Thus personnel procurement and development alone is not sufficient. Among the number of potential services travel agencies must identify those that set them apart from the rest of the competition. These services must be subject to investment. In order to avoid bad investments, the core competencies must be linked to the market's demands. By means of market oriented core competencies it is ultimately possible to master the strategic challenges.

6 Conclusion

The business travel market is very dynamic. Barely a year passes without incisive changes. Service providers, travel agents and travel managers must constantly meet new challenges. In the future some of the central issues will be:

- Further development of sustainable technologies for SBTs, quality control systems and MIS;

- The transition of travel management companies from mere booking agencies to extensive advisory firms;

- Eco-friendly business travelling;

- Security aspects: protection of business travellers from harm (terrorism; natural catastrophes);

- Fair consideration of the interests of service providers, travel agents and business travellers.

Welcome to Adventure Land

Chances and Risks in Internationally Oriented Travel Management

Jörg Martin

Phelias Fogg, the protagonist in Jules Verne's "Around the World in Eighty Days", is a master of efficient travel. And it's no wonder, as his trip is simply a means to an end – to keep his fortune out of the hands of his London betting mate. Consequently, Phelias Fogg is a business traveller par excellence. And as such he experiences, much as the business travellers of today, that the world outside his own offers many surprises, unexpected obstacles and all sorts of temptations. Of course, Phelias has the advantage of being a fictional character, dependably striving to achieve his literary purpose – a happy ending. This is not something guaranteed in real life, and particularly not in travel. In fact, travel is the Achilles' heel of many businesses. It is a sort of black hole that irresistibly attracts and subsequently consumes money and resources. And it continues to do such until cost control or management hits it with the "E" question – is it efficient? If the question has not been posed previously, it quickly will be when the company goes international and travel volume and organisational requirements begin to grow. Now is the time for process design; now is the time for travel management.

Travel management goes international. Due to advancing globalisation, the subject is gaining priority not just quantitatively, but also qualitatively. Who should work this field and what tools should they use? What goals should be focused on and where are the starting points? These are questions that many businesses are having trouble answering, often because they lack experience with travel management concepts in their domestic market. Leading businesses from English-speaking countries and other major corporations with the necessary management structures have already approached the subject. It is important to note here that the methods used for internationalising travel management are heavily influenced by the culture of a business's domestic market. American businesses, for example, operate very confidently and with clear directives. European companies tend to operate more cautiously and prefer integrative persuasion methods.

Taming the Shrew

Regardless of the culture travel management is based upon, all concepts strive to unify internal and external travel processes beyond national borders. Indeed this is a Herculean task that requires nothing less than consolidation under specific and to the outside observer, often enough, very strange foreign market conditions, which tend to be impervious to order.

Nevertheless, it is worth the effort, if only because the potential for optimisation is substantial. And not just in the international dimension. The benefits businesses gain from optimising travel behaviour in foreign markets also work in their region. The greatest rewards will be reaped in the domestic market.

What Are the Requirements for Achieving Internationalisation?

National Consolidation Has Precedence

Before international consolidation becomes an item on the agenda, national travel management should be in top form. Only when efficient processes have been established in the domestic market and cost reduction potential therein exhausted, is it recommendable to start looking across national boundaries. A respectable showcase and a good success story are critical to be convincing beyond the border.

Identify Responsibilities

Few businesses have the international network structures required to unify processes. There is a simple reason for this: travel management has varying organisational foundations in individual markets. For example, in Germany, travel management is often the responsibility of the purchasing or human resources department – that may not be the case in France or Brazil. This means that international consolidation will first require that those responsible for the topic be ascertained for each market and business, in order to acquire the necessary information about the current state of travel management in a particular country.

Never Without a Clear Mandate

The most important requirement for national and international process optimisation is a flexible mandate from company management. Consolidation cannot be accomplished without this mandate – neither at home nor abroad. Unfortunately, this aspect is frequently underestimated, even by project representatives. When asked at travel manager workshops what their mandate and commission are, project representatives are often unable to provide a spontaneous answer. This leads one to conclude that the commission is not official, but, rather accepted. It is often confirmed that a flexible mandate does not exist. This means that the business has not thoroughly defined the field of responsibility. What does the business expect from travel management? The answer to this question should be as clearly formed as possible if travel management is to have any chance for success. When in doubt, the travel manager should pick up the commission and start asking important questions. What does my employer expect me to accomplish? What are my goals and what room for manoeuvre does travel management have?

The Chiefs Need to Be Convinced

Internationalising travel management is a multifaceted task as exciting as globalisation itself. It is a task that offers substantial optimisation potential, but also involves risks that cannot be ignored.

One of these risks is encountered by every travel manager in the form of powerful regional leadership. If travel management is not equipped with a resilient mandate, conflicts are bound to arise. The pack leader assumes an ostentatious pose and tells management at home that they will not accept any intrusion into their area of responsibility and that everything is functioning fine anyway. Should support from the chiefs still be withheld, the project will die before it ever gets started. Conclusion: internationalisation should be based upon a success story in the domestic market and have the blessing of the highest levels of leadership.

Internationalisation Is Not a Desk Job

After the mandate is secured, the commission is defined and those currently responsible for the field have been identified in each market it is time for the first analysis. This usually entails transparency of costs and processes for the market in question, for example, by using a process-check form that enquires about volumes and transactions by category, travel guidelines, process flows and supplier contracts, thus providing first indications about the status of travel management in a given country. The groundwork is concluded with a raw concept that, above all else, prioritises the countries in question.

Internationalisation cannot be accomplished from behind a desk. Travel management must be achieved within the individual markets, in order to gain the acceptance and participation of the people affected. This is the only way to create any leeway for the arrangements that are made – particularly in countries with a different cultural background. It is very important that the project representative, their supervisor and the managing director of the market in question all participate in the first meetings. Ultimately, it will come down to the meat of the matter – mediation of the commission, strategy, validation and supplementation of the elevated market status, identification of market specifics and, of course, solidification of approach methods. Naturally, a detailed protocol of this kick-off meeting will be created and passed on to the participants and management.

Obviously, consolidation will require constant contact with the individual markets. Quarterly reports and status reports are suitable media to integrate the people responsible at location into the project and to commit them to success.

The Reality Check

The potentials of internationalising travel management can be seen more clearly by observing real world travel. Following the creed "nothing is impossible" tickets are rebooked on location for an airline other than the one prescribed, a different hotel is checked into – naturally at the employer's cost – and a down day is squeezed in, because flying business class is so strenuous. Travel life is fun and, for example, switching to an unauthorised hotel to collect incentives, such as bonus miles, it is not uncommon. Hotels report regularly about how travellers pressure reception desk personnel into giving them a room upgrade. And, of course, commercial exchanges during business trips often include the oldest trade in the world and the discretion proffered is almost never kept. Particularly in China conflicts often arise when travellers escorting prostitutes to their rooms disregard that prostitution is illegal in China. In short, bending or even breaking the rules of travel is commonplace and not just when travelling abroad. Since everyone simply does what they want, travel management remains, in reality, often just an illusion.

The behaviour of many travellers is not just questionable from a moral standpoint, but also directly influences travel management efficiency. For example, when negotiating hotel rates the reputation of a business always plays a deciding role. Or, put differently: companies whose employees know how to conduct themselves have a definite advantage.

Accordingly, sustainable travel management requires facing unpleasant, and often enough unsavoury, facts. Only those who know how it is can determine what and how it should be. It will quickly become clear that the conduct of travellers directly reflects upon the current state of travel management. The system will be abused wherever it can be.

The discrepancy between requirements and reality, i.e. between a standing policy and the actual practice, can only be resolved when the guidelines are thoroughly communicated and enforced. Travellers must be included in a network, in which qualified information is provided centrally and is easily accessible. The advantage is self-evident. Unified process control, easy access and transparency all have immediate impacts on cost-effectiveness.

Establish Structures

In order to define valid guidelines for all markets, to install uniform processes and establish an efficient communications infrastructure, resources will need to be pooled. The travel office's account management plays a deciding role in this. The best proven systems so far are those in which a key accountant in the domestic market manages communication with the diverse travel office partners worldwide, gathers relevant data from the markets and establishes quality control. The strong position that a business usually has with its domestic partners can be an effective tool in this process. Their position can also be useful in international activities as sales volumes in core markets can make for a good argument when negotiating preferable conditions in foreign markets. Additionally, the goal of minimising process, travel office and primary travel costs is incumbent upon travel management.

Who Is Allowed What – Service Standards

The idea that everyone is the same is an illusion that will confront travel management no later than when the question of developing uniform service standards arises. If for no other reason than simplicity, it would be desirable to treat all employees equally, regardless of varying origins or hierarchy levels. However, service standards will be opposed by the varying cultural customs of the markets – or even the, more or less clear, expectations of the business's management body. This may lead to expecting Chinese or Indian co-workers to fly economy class on long-distance flights even if business class is the only way to fly in the domestic market. A hotel below the envisioned standard will be reserved for an Asian colleague or colleagues travelling together will be allowed to book separate hotels of higher or lower quality based on "social status" – in reality an absolute "no-go".

The decision as to how differentiated the travel guidelines should be, will be made by the business's management body based upon cultural, social and political considerations. Of course, travel management, with its expertise and grasp for local conditions and emotional compositions can and should take part in this difficult discussion on company policy. Ultimately, the main task for travel management will be to convert specifications – no matter how problematic they might be – into binding standards, to clearly communicate the rules and to oversee their enforcement.

It is important to note that uniform service standards do not necessarily reduce costs. Polishing estates is not the main concern. What is important is the establishment of a transparent, comprehensible and balanced policy as a fundamental component of optimised, controllable and therewith overall more cost-effective processes.

Standards in this form can even become a burden, as the pack leaders will quickly point out if they feel their profit shares are in danger. All-in-all, at least at the international level, travel managers must negotiate a minefield. Travel managers require diplomatic skills, compelling arguments and the necessary support of their superiors.

Do Good and Spread the Word

Those that issue the mandate are often impatient and want to be kept happy. Thanks to a consolidated database, efficient process designs and a well conceived and communicated policy there should be no reason for a travel manager to need to hide in a corner. Quite the opposite is true. They should be telling upper-management what they want to hear, for example, that the travel budget can be cut by 25 per cent within a foreseeable period. Intensive reporting based on consolidated data makes travel management operations transparent and provides the department with standing that is above and beyond the mandate and cannot be ignored by change management.

Intelligent Travel

Any company that has ventured into globalisation has, more or less painfully, learned about foreign worlds. The ability to adapt is critical. Travel management can often play a key role in providing the required adaptation. Ideally, travel management will be able to make use of local features and turn disadvantages to their favour – an example:

Tickets for flights within India can, for some time now, no longer be purchased with a foreign credit card over the internet. This is because Indian officials decreed such in an unfathomable resolution. Well developed travel management would have a remedy. They would have already acquired a long standing relationship with a travel office in India that could book the tickets locally and then forward the electronic tickets to the domestic office. It is a simple arrangement that will save a traveller the hassle of finding their airline's ticket counter in the Delhi Airport late at night, in order to pick up tickets to their destination.

Or, another example – this time with a direct cost effect: An internationally based business has its American headquarters in Boston and its German centre in Heidelberg. Communication between the American and German travel offices functions smoothly. And the travel office representative in Boston knows that they are better off buying tickets in Germany, since the price difference between the USA and Germany is striking. Considerable cost advantages can be realised in this manner.

Lost in Action

Since the conditions in foreign markets are sometimes drastically different than those at home, travel management often provides a form preventive crisis management. Travellers are constantly being confronted with problematic conditions in their destination countries. It is important to be prepared for all contingencies ahead of time and to structure plans in such a way as to ensure that the company does not neglect their responsibilities towards their employees. The traveller must be given clear guidelines as to how they are to conduct themselves and who is to be their contact, for example, in case of an unplanned change of location or a medical emergency. The worst case scenario is one in which the company loses contact with the employee. This will almost certainly occur whenever a traveller books flights on their own without consulting the company. Solid supplier relationships in the markets, corresponding access to data and clearly defined conduct guidelines are essential to ensure that this kind of situation does not arise and that the company can respond appropriately in case of an emergency.

Many Settings, Many Risks

Risks for internationally based travel management wait in very different settings – at the very least because globalisation has yet to lead to harmony in technical processes or judicial standards. Even those that work with international providers will quickly run into obstacles that will require, among other things, a degree of improvisation skills to overcome. An example with travel offices: agencies in the individual markets work with very different GDS systems – in Germany with Amadeus, in the USA with Sabre and in Russia and China with many varying and not inter-compatible systems. In China alone there are at least four GDS systems currently being used by airlines. The systems we are accustomed to are not allowed to be used there. This creates a problem for travel management back in the domestic market, as reporting is largely based on system data. The only available option is improvisation. The data can be entered manually in the domestic system and then deleted in the original. This is, however, not exactly a model for efficiency.

Travel management in Russia can also be particularly troublesome. There are currently five reservation systems in use in Russia and none of them go by the name of Amadeus, Sabre or Galileo. A majority of the travel offices are run by franchise businesses that are seldom accessible for warranty claims. Due to the completely different conditions in Russia, it is all but impossible to maintain a desirable quality of uniform reporting.

Another similar problem in Russia involves the subject "back office". Russian travel offices commonly provide just one ticket copy, much to the joy of accounting. Or a credit receipt will arrive in which the line "Invoice" has been struck through and replaced with "Refund" – of course, with the original receipt number and a positive sum. Again, accounting is thrilled while the travel manager attempts to explain the basics of western European accounting standards to the Russian travel office representative.

In general, no other country has as many bizarre surprises in store as Russia does. Anyone who has ever dealt with Russian airlines will have some sense of this. Take an illustrious airline, such as KrasAir. A traveller will find them anything but charming when winter sets in and runways are covered in snow.

Summary: The established quality standards in Western Europe are seldom found in developing countries. India is an exception. The country is in some respects more advanced than Germany when it comes to the travel office sector, for example, with regard to automation, service and automatic tracking of internet-based visa procurement processes.

Hurdles in Reporting

Uniform and substantial reporting lies at the very heart of qualified travel management. It provides the transparency that the mandate issuer expects and is indispensable in regulating provisions. Consistency and quality are deciding factors in reporting. And therein lies the problem. The standards and practices in the various markets will lead to very different outputs. Whereas in one country gross costs are emphasised, a partner in another country may provide ticket data or net costs. If the system is not unified at the lowest common denominator, inconsistencies in the data will lead to a distorted perception and consequently to the wrong conclusions. Anyone who wishes to avoid these stumbling blocks would be well advised to constantly question the processes and data sources.

Create Synergies

In every risk there is an opportunity. Of course, this goes without saying. But it is without doubt valid in travel management. Imagine for a second that business "X Worldwide" works together with the globally present travel office chain "Team Travel". The co-workers here book locally, as do those in Canada – good business for the agency. But amazingly Team Travel acts as if there were two distinct companies at play here. For example, if a Canadian colleague flies to China with a stopover in Germany to visit the centre there, the rates negotiated in Germany will not be applicable in Canada as well. This is solely due to lack of communication between the travel office representatives on both sides of the Atlantic and exchange between the markets. The employees in Canada believe that their offered rate between Germany and China is good, even if they pay 5,000 Euros for a business class ticket to China and the current German rate is 2,000 Euros. Communication is key – if only to take advantage of the enormous potential offered by a truly global arrangement. It is necessary to ensure that rates are compared and travel office partners are involved with the key accountant on location.

The Human Factor

Ultimately, friction for internationally oriented travel management will rest on the shoulders of those involved. Because, even though everyone knows that we now live in a globalised world, local actions are still stuck within the borders of "This is what we have always done." This holds true for Germany, Canada and particularly for China. In China it is not uncommon that a travel office's account management is responsible only for a certain, very constricted region. Anyone who wishes to get their Chinese partner to cover the entire market will be met with amazed faces, since no one there has yet to contemplate such a notion. In reality, globalisation connects many villages. Not much more has been accomplished yet.

Even within a single culture group there are often enough different uses and habits or simple misunderstandings to provide human challenges to travel management. Speaking to an American partner might take some extra effort before they understand what is expected of them – and there are no assurances that the message will be truly understood. "Yes we can" does not always mean "We will get it done, James."

It should be clear by this point that travel management requires resources – mostly time and money. Telephone conferences and e-mail alone will not suffice to achieve an acceptable result. Those who wish to take full advantage of the enormous cost reduction potential will need to continually immerse themselves in the markets and keep in close contact with the appropriate people.

Sometimes It Is Better to Be Lucky than Good

Sometimes talking will not get the job done. The only help left is luck. For example, when despite global requirements and escalation, access to the local agencies of global partners is simply not possible or when existing structures exclude the possibility of cooperation. These phenomena are observed time and again when the partners on location are simply franchise participants. This often the case with travel offices and credit card institutes. The total volume indices that inspire businesses will receive nothing more than a tired smile from the franchised dealer, since they don't get anything out of it.

The same difficulties intensify a different problem when working with globally active partners – the problem of dependency. A travel office or credit card partnership that has been implemented at great cost cannot be simply thrown out the window, especially if obstacles in the market against the unification had to be overcome. After all, credibility is at stake here. Of course, the partner knows that. And you are aware that the laboriously constructed international structure is very fragile, particularly in its early stages. When travel management is still young, and requires extra care and attention, it is almost impossible to switch providers, since a new partner will certainly not be found soon enough. Put plainly: a change of partners in the early phases is suicide.

It is equally important to keep a close eye on the work done by partners and to let the international project grow in the interim until the processes have settled in. After about four or five years the structure should have stabilised and travel management can be expected to have gained standing. At this point a change of partners can be considered, if the benefits are great enough.

Know Who Is Committed

It is important to be careful when selecting partners. Not every provider is prepared to accept domestic market volumes at foreign market rates. In particular American airlines are adamant about this point. Even a good contract in the domestic market as a strong argument might not necessarily help. Companies in America will insist upon their own, high entry volume quotas of up to a half million dollars and will on top that ask for travel management data material in order to estimate expectable volumes – something that it absolutely unheard of in Germany.

So global does not automatically mean cheaper and most definitely not better. Instead of relying on synergies that due to the company politics of the potential partner may never be reached, travel management must balance what and how much should be done with which partner, if for no other reason than to avoid the risk of becoming a victim of globalisation instead of profiting from it.

Real globalisation will require a shift in the awareness of providers that is far from being realised. Nevertheless some are setting examples, such as Air Berlin. The argument "one company, one rate" has run into refreshing good will there; one reason why many businesses are using Air Berlin to fly around the world at uniform rates.

Return of the Chief – and His Mates

The agenda has been executed, structures have been put in place, the right partner has been found and processes have been communicated. Certainly nothing can go wrong now. Yeah, right! As mentioned from the very start, travel management and its internationalising project will run into obstacles within its own organisation time and again. And in the fight against change, money is a compelling argument. Thus it may come to pass that the chief in India sends an agitated mail asking why they have to pay part of their flight cost, when they have always flown for free in the past. This is a typical example of an Indian relationship and dependency network against which European cronyism seems almost ideal. Accordingly, the efforts needed to ensure that the agreements met are actually put into operation can be demanding. A one year visa and an appropriate stock of charcoal tablets are definitely worthwhile. On top of everything else, infringement on long-standing relationships is a balancing act that does not make the issue any easier to deal with.

Of course, travel management does have the better arguments: uniformity, transparency, crisis security, cost effectiveness, controllability and equity. The more the person responsible for the project knows about the conventions of the individual markets, the more they understand the dos and don'ts, the better the chances are that the right arguments can have a positive impact even under the most exotic conditions. Exchanges with colleagues that know their way around the jungle of local conditions are of utmost importance. Another way to prepare to work in foreign markets is to visit a relevant country workshop, such as those offered by international commerce chambers.

Often Overlooked Problems

Travel guidelines are all well and good. They can be real cost killers, as long as they are followed. Travel management would be well advised to close any organisation or communication gaps that could allow travellers to get around the rules. This too cannot be accomplished from behind a desk. For example, when a German co-worker with a mission in Delhi regularly coerces the Indian colleague responsible for booking local hotels into booking outside the guidelines, the only thing that can help is a discussion on location. The goal is to overcome the habitually submissive demeanour that Indians tend to display towards Germans, so that they can say "no" at the deciding moment. After all there is a lot of money at stake. This holds particularly true in India, where the price differences between hotels can be so exorbitant that even executives of large German companies pay close attention to price lists when travelling there.

The coherence of processes can be equally critical in deciding outcomes. Put differently: if there are any hiccups in the system, even the best travel guidelines are useless. They will simply be ignored. A real world example: a travel group is supposed fly from Germany via Moscow to the Asian part of Russia. Their travel guidelines limit them to economy class for flights within Europe. Business class is allowed for intercontinental flights. The attentive colleagues believe it is to their advantage to book their flight from Moscow over the Urals not in the German travel office, but instead through a representative in Russia. The domestic travel office books the Moscow flight, and in business class at that, since the group will continue on to Asia. As it later turns out, the trip ends in Moscow for our industrious expense account fiddlers. They can only be complimented. And then the search for the error in the system will begin.

In the example above the travel group was only able to release their business-damaging creativity, because the problem with the inner-Russian flight was not covered by the processes. Ideally the appropriate local travel office would have contacted a Russian partner agency for the booking, so that all the strings would lead back to a central point.

Praise for Baby Steps

One of the most obvious consequences to be derived from "foreseeable" implementation difficulties is that markets should be dealt with successively. In addition to reason of capacity, this approach method is supported by the experience that many problem areas are similar and proven solutions in one market can often be conveyed to other markets.

A further prerequisite for successful implementation in foreign markets is the installation of a single point of contact for the international organisation – one body that is responsible for travel management on-site and can ensure that booking data are available in close to real time by controlling the local provider in co-ordination with the key account manager. Without a reliable data flow the system will remain impotent and the situations in the markets will remain as chaotic as they were before.

Speaking of data, every travel management solution, whether in the national or international dimension, is, of course, IT-based. The design of the IT architecture is as specific as the corresponding business requirements and the reporting demands. However, there are still a few basic rules that should be taken to heart. One is the principle "no ticket without a project". This means that every trip will be assigned a project number by the travel office. No ticket will go out without one. As trite as this may seem, it drastically relieves the burden for the entire process, all the way down to finance accounting. Good travel management will always keep process costs in mind.

Always remember people want to be convinced and entrained. It is important that internationally oriented travel management does not come off as a repressive regime, but instead as a cultural project that is useful to everyone and is in line with the company's identity and leadership. Travel management needs to be seen as a project that noticeably improves service quality for travellers, assures maximum personal security in problematic markets and creates transparency where the dark mists of hierarchy and unquestionable privileges once swirled.

When all is said and done, travel management has a highly political dimension, at least when it is internationally oriented. It is completely and fully bound to the spheres of real-world economics. It does not just blow smoke, but instead creates tangible cost advantages – a benefit that companies come to embrace, particularly in difficult times.

Multicultural Interactions During Meetings and Events

Culture Eats Processes for Lunch

Uwe Klapka

The question of whether the humble boiled egg should be hard or soft-boiled can be a difficult one to crack, so to speak. Everyone has their own preference. The relationship between men and women, too, is fraught with division; they simply do not belong together. German comedian Loriot, alias Vicco von Bülow, has a sketch in which he skilfully brings these two difficult subjects together as an elderly married couple argue about the art of boiling an egg. What Loriot so brilliantly manages to encapsulate in this satirical sketch, is the lowest problematic denominator of intercultural difference.

100,000 years ago human beings made their first tentative efforts at communication through words, sentences, rudimentary language. God, says the Bible, punished the hubris of those engaged in constructing the Tower of Babel, with the confusion of tongues. And nowadays, no matter which language we speak, what we say to one another is very often misunderstood.

The origins of Western modernity lie in economic globalisation: Marco Polo, Jacob Fugger and Christopher Columbus went out into the world to do business. They were searching for raw materials and goods, trade routes and markets, what they encountered were people living different kinds of lives and talking in different languages; new, strange cultures. The lives of those early explorers meant that they travelled for months, even years, at a time, their entire lives dedicated to travel. The pace of modern globalisation is rather different. And to be successful, the process no longer relies on subjugation and colonisation, but rather on participation and communication. *Shift happens*! – change is a constant, even, and perhaps especially, in the constantly growing meetings and events industry. It's true to say that there is nothing new in this, but is the message being heeded?

- The global village with its growing infrastructure and contact networks already exists

- Multinational organisations are competing for high potentials (The War for Talents)

- There are more and more reasons, an increasing need, for international meetings and events

You may well feel that you've had your fill of platitudes on globalisation. And you'd be absolutely right!

But we are NOT talking here about the Englishman who means what he says, but does not say what he means ("very interesting"). Nor about the Finnish Swede, who can be silent in two languages. And this has nothing to do either with the question of how long a German should study the business card of his Japanese associate.

What we are talking about is YOU. About your relationships to colleagues, fellow students, friends, business partners and loved ones. It's about your (and my) ability to communicate, with one person or with several, with people you know or with strangers, and to do so as clearly and unambiguously as possible. Admittedly, not always easy. What happens, however, if we (and this is what we want, after all) as members of some random group at a meeting want to (or have to) communicate or to participate in discussion?

Successful communication with people from other cultures demands above all that we understand the particular values and qualities of those cultures. This is not just about national cultures; there are plenty other cultural resources; ethos and religion, for example, sex and generation, education and profession, or social and economic position (which others can you think of?). Cultural behaviour arises from accumulated experience and cumulative knowledge, filtered by and transmitted over many generations.

Every culture is composed of a multitude of highly complex components, of which we are capable of perceiving only a fraction. Let's think for a moment of the famous iceberg. We see only the tip: arts, food, mass media, fashion. But hidden below the water line are all of those components that offer potential for conflict and a high degree of cultural misunderstanding – social norms, rituals and taboos, communication structures, speech styles and listening habits, audience expectations, non-verbal communication, the use of time and space (interpersonal distance, eye contact, silence), and, last but not least, ethical values (national characteristics, religion, attitudes and world views).

Meeting Professionals International (MPI), the world's largest association in the meetings and events industry, publishes an annual international study of the events market, Future Watch, which is published in collaboration with American Express. In the most recent issue, 60% (1600) of the corporate planners and suppliers surveyed said that they believed their business would expand globally in 2008. Not only multinational employees were of this opinion; owners of small and medium-sized enterprises were also in agreement. Only 1% of those questioned expected any decline in globalisation (Future Watch 2008 – MPI).

The impact of culture on the global economy (and, of course, on meetings and events) continues to grow – whether we like it or not. A law of nature, so to speak.

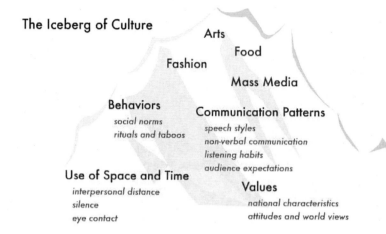

Fig. 1. The iceberg of culture

Why is it then that we (you and I) as representatives of the global economy have such a hard time coping with the consequences?

Let's take a look at some of the practical impacts of the intercultural spoilsports.

International mergers have increased dramatically in Europe. However, 83% of mergers produce no added value (source: KPMG homepage). On the contrary. Mergers often bring rather considerable operational and economic damage in their wake.

Remember the marriage and divorce of DaimlerChrysler? Or what about what happened to Wal-Mart? The group lost 1 billion US dollars (and hundreds of jobs were first created and then destroyed), because the American bosses simply didn't realise that we Germans cannot stand to be confronted at 10 in the morning by an employee, mouth fixed in a combative rictus grin, who insists on packing our shopping bags for us (They're MY bags!).

There is no doubt that a strategy such as this would have been prepared after a great deal of data gathering, project planning and careful process and change management. Nevertheless it failed, and it did so because it neglected to take cultural differences into account. Culture eats processes for lunch.

Is it not surprising that 72 per cent of M & A managers worldwide admit to taking no account of culture-specific considerations in their business decisions (Hayes Report 2007)?

Much still has to be done! Where, however, are we to find the means and the methods to put into practice in intercultural business life?

I want to mention the name Richard D. Lewis to you. For the benefit of those of you unfamiliar with the name, the 78-year-old Englishman is one of the world's most renowned philologists and linguists. He speaks ten European and two Asian languages. His pupils have included a host of famous names, including Sean Connery, former footballer Gary Lineker, an ex-president of Finland and even Empress Michiko Shoda of Japan (along with several other members of the Japanese royal family). Knighted in Finland (yes, really!), Richard D. Lewis is author of the best-selling "When Cultures Collide".

He has spent his life exploring the influence of culture on individuals, social groupings and businesses. His "Richard Lewis Model of Culture" is a highly complex, theoretical and yet practical model that breaks multicultural interaction down into its component parts and gives them the kind of clear classification that makes them eminently suitable for use in the multinational business environment.

As business activities become more international, more global, the demands on intercultural skills, too, increase proportionately. It's not easy to get your business message across and convince people with the same background as yourself. How much greater is the challenge then when one is dealing with people who have different values, perhaps even different gods, and who organise their world in very different ways. How is it possible to create a productive working basis in such an environment, and one that is going to keep everyone happy?

But it is possible! And what's more, different perspectives on products, processes or problems make it possible not only to progress economically but also to do so in areas such as ecology, science and philosophy – like the beginning of the Enlightenment. And yet it moves!

It all began with a multicultural initiative by MPI in 2001, supported by Disney World. There were good reasons for this. MPI brings together 24,000 members from 80 different countries and thus provides a meeting place for many different languages and cultures. The aim of the initiative was to establish what sort of cultural barriers our members identified as things that needed to be considered when planning events.

This work led to the production of a research study published in 2005/06 as well as to the development of the "MPI Meeting Culture Toolkit" (150 pages). This tool is intended as an aid to members and other people involved in the meeting industry. It aims to help them both to overcome cultural barriers and learn to understand, accept and make productive use of such barriers.

A total of 1,743 members, aged between 21 and 75, from 61 countries participated in the study. 76% of respondents were women. The study consisted of open-ended rather than standardised questions.

The study included the following questions:

- What does the term multicultural mean to you?

- How can organisations and businesses better serve multicultural markets?

- How would you define a successful meeting?

- What are the greatest deficiencies in how organisations meet multicultural needs?

More than 90% felt that businesses and organisations needed to become more sensitive to culture and ethnicity issues. MPI members identified the following points as important for the success of multicultural events:

- Awareness and recognition of social cultures and awareness of practical issues such as dietary requirements

- Increased cultural education and special training within companies

- Understanding of regional and international cultural differences were seen as very important

- Multilingual staff

The co-operation with Richard Lewis resulted in the creation of the web-based e-learning tool "Culture Archive".

The "Richard Lewis Model of Culture" classifies world cultures into three categories: multi-active, linear-active and reactive. Lewis then assigns what he sees as their essential characteristics to the three categories.

Family, hierarchy, relationships, emotions, eloquence, persuasion and loyalty are the chief components of the multi-active category. Main characteristics of the linear-active category include facts, planning, rules, law, while the key features of the reactive category are intuition, harmony and reliability.

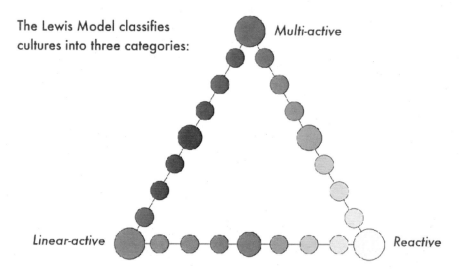

Fig. 2. Richard Lewis model of culture. Source: www.mpiweb.org

The typical representative of the multi-active type is loquacious and reacts emotionally. He tends to be rather mercurial and impulsive in thought and act, but is nevertheless very productive.

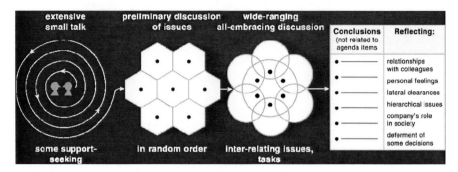

Fig. 3. The multi-active type. Source: www.mpiweb.org

Linear-active types have less to say for themselves, but are more direct, stick to facts, like to work according to plan and put truth before diplomacy.

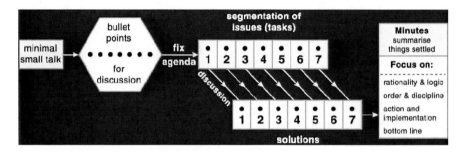

Fig. 4. The linear-active type. Source: www.mpiweb.org

The reactive person is a good listener, polite, but indirect, avoids confrontation and tends to put diplomacy before honesty.

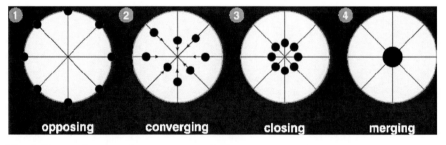

Fig. 5. The reactive type. Source: www.mpiweb.org

The Lewis model shows what happens in meetings where representatives of all three categories are present.

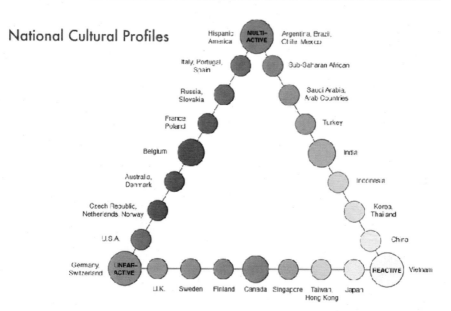

Fig. 6. National cultural profiles. Source: www.mpiweb.org

Of course it is true that individuals, whichever culture they belong to, do not lend themselves to unambiguous categorisation. So the model includes a multitude of hybrid forms that span the areas of ambiguity and variation between the basic categories. Only then does Lewis assign national provenances to the resulting cultural profiles.

In collaboration with Richard Lewis, MPI has turned this theoretical model into a practical web-based learning resource for their 24,000 members. In a first step, personal verbal and non-verbal skills are analysed in the Personal Cultural Profile before being assigned to one of the hybrid cultural groups. People or social groups can then relate their (inter-)cultural skills and temperaments to one another.

The user initially ascertains his/her own inter-cultural profile, which is determined by an assessment centre. Finally, there is an opportunity for users to compare themselves or their groups with other multicultural groups or colleagues.

There are currently almost 100 national profiles available which can be compared with one's own individual behaviour. The resulting report provides advice on where and how one can better adapt oneself, and what areas still need to be worked on. When in a reactive culture such as the Japanese, for example, it is important, especially for multi- and linear-active people, to keep expressions of emotion and gestures to a minimum. The "MPI Culture Active" tool also provides important lessons for the planning and organisation of meetings and events, which can be quickly and clearly implemented within the context of meetings with other culture groups. It also describes appropriate business conduct for the benefit of participants in multicultural meetings.

Whether, for example, one chooses one's own or another country, one immediately receives various pieces of topical information and data, which is updated on a daily basis. The user receives information on how events are conducted in the various countries, and which of the listed details are of particular importance, as well as information on religion, the political situation, geographical conditions, history, politics, the economy, cultural interactions, languages, basic beliefs, culture, body language and communication structures. All of this is then broken down to provide information that is relevant to the planning of meetings, conferences and events in the respective countries.

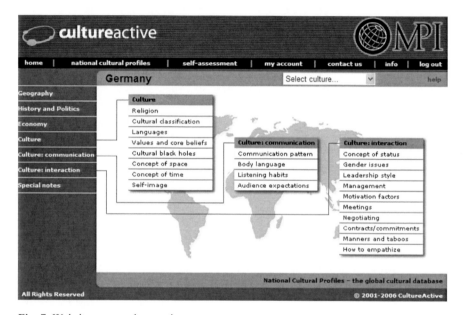

Fig. 7. Website www.cultureactive.com

You still can't quite see how all of this works yet? Send me an email and you'll be given trial access to our MPI tool. Take the fun approach to getting to know what our 24,000 members from 80 countries already know and appreciate.

And, by the way, I myself work in MPI's European headquarters alongside ten colleagues from 8 different countries. We don't always understand one another straight away. But for us this diversity is an enrichment. The broadening of cultural horizons that such a multinational team generates brings with it an enormous increase in creativity and ideas.

A strong sense of humour helps us overcome all language and cultural barriers together. If difference is to be productive one must be able to laugh – even at oneself.

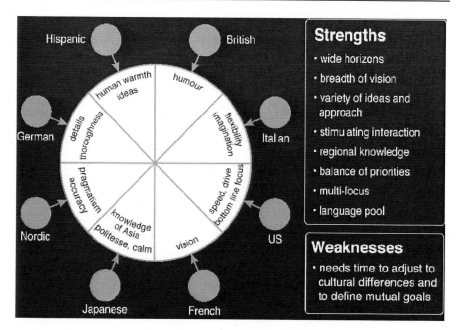

Fig. 8. Cultural characteristics and collaboration. Source: www.mpiweb.org

P.S. Have you heard the story about the young German coastguard who received an SOS call from a sinking English ship in the middle of his first night on duty? "Mayday! Mayday! We are sinking! We are sinking!" comes the message over his radio. "What are you sinking about?" replies the coastguard.

Security on Business Trips – A New Task for Corporate Travel Management?

Andreas Krugmann

1   An Employer's Legal Obligation to Protect Employees

Do companies think about whether their employees are adequately insured when they go on business trips? Do they think about how globalised the business world is and how uncertain conditions are in many countries nowadays? Far from it; when it comes to the way many companies view the risks their employees face during business travel, the prevailing attitude appears to be a pithy, "Nothing's happened yet, so nothing probably will happen." Even business travellers themselves — no matter whether it's a senior manager or a construction worker under contract — do not seem to be aware of potential problems. The only time the topic of travel insurance arises is when it comes time to go on family holiday. According to security experts, many people seem to think they were magically protected by their employer while on business trips.

To a certain extent, this is a valid assumption, since a work contract gives employers a variety of responsibilities in terms of meeting the needs of their employees. In Germany, for example, paragraph 618 of the civil code expressly states that life and health are objects to be protected. The related obligations implicit in a labour contract mean that employers must do their best to protect employee health and wellbeing. Doing so, however, is vastly more complicated when a construction project has to be carried out in a troubled area overseas than it is during a business trip in a neighbouring European country. Are people generally too lackadaisical when it comes to business trips?

The sheer volume of business travel alone reflects the need to place greater emphasis on security whilst travelling. According to the 2007 analysis of business travel conducted by The German Business Travel Association *(Verband Deutsches Reisemanagement [VDR])*, business trips totalled €48.7 billion — only slightly less than the amount spent on holiday travel. Small and mid-sized companies with 10 to 250 employees made up over 50% of this total sum.

These smaller and mid-sized companies frequently do not have adequate insurance coverage for their business travellers. Many firms are convinced that their employees are taken care of during business trips thanks to a combination of personal insurance, professional associations and credit cards. The rude awakening comes when an incident occurs and the company has to pay for the damage.

The main factors that cause employee downtime on business trips are differences in climate, failure to adjust to jet lag, job-related stress or foreign food. There are also many other risks which will be discussed in greater details later.

Many companies are often unclear about the expenses which may result if, for example, an employee needs hospital treatment overseas. In the US, an operation and intensive care after a heart attack can easily amount to a large six-figure sum. Major expenses may also occur if an employee has to be brought home after an accident in a country where health care is substandard.

Incidents can even take place during domestic trips, jeopardizing the success of the business trip and incurring costs. Stolen notebooks and lost travel documents are only two of many examples.

2 Discrepancies Between Employees and Employers – How Do Travel Managers Assess Security on Business Trips?

"German companies neglect travel medicine. The extent to which German companies offer their employees information about travel medicine is sobering. According to their own statements, 77.3% of employers do not talk with their employees at all about health risks that are present during business trips at holiday destinations. 12.5% only offer this information to their business travellers. Travel medicine, however, is growing in importance within the field of corporate risk management, since there is a great deal of potential jeopardy from imported infectious diseases, such as a flu epidemic. The gaps that exist in preventive measures at companies are made more intense by the people's lack of knowledge: 41.7% of travellers are not aware of the potential risk of disease at holiday sites. 18.5% have said they intend to get informed before their trip begins."

This was the result of an interesting 2007 study held by the Hamburg-based research institute *Institut für Management- und Wirtschaftsforschung* and handelsblatt.com on the risks of falling ill overseas. Tbilisi today, Quito tomorrow, Shenzhen the day after – with travel itineraries like this, whether or not someone looks the part for a business meeting is not nearly as important as making sure employees are prepared with proper information.

In the fall of 2006, the first part of a study was held to examine the subjective sense of security that travel managers have versus the reality of situations during business trips. A 2007 study, the ELVIA security barometer on business travel, was conducted in conjunction with the University of Lüneburg.

The study found an enormous discrepancy between the subjective feeling of security among the travel managers polled and the actual state of things during business travel. When it comes to employee security on trips, the findings in German companies are very divergent. According to 76.5% of travel managers who are responsible for organizing employee travel, the majority of business travellers feels secure. However, upon asking the travel managers specifically whether or not an unanticipated problem had ever occurred, 79% said yes.

When looking closely at individual risk scenarios, one also finds major discrepancies between people's feeling of security and the occurrence of actual problems. Lesser risks such as theft or lost luggage are anticipated more often than they truly occur. Kidnapping, death, unrest, natural catastrophes and attacks are issues that the surveyed travel managers tend to overlook altogether, and they take place more often than assumed. In comparison to the severity of the overall risk, business travellers are frequently confronted with problems such as kidnapping (4%), upheaval in the host country (19%), natural disasters (23%) or attacks (22%) (see table). The startling finding is that only some 40% of all companies have a response plan for such contingencies. Only a third of travel managers cite accidents as the top three most common travel risks, although accidents have happened during a business trip in over half of the companies surveyed. If nothing else, 60% of companies are prepared for this scenario.

The reality of the situation, however, is that employees are affected relatively often, and it bears repeating that contingency plans exist in only 40% of companies. Apparently there is little awareness of the potential dangers, which is due in part to the fact that the only situation that makes headlines is spectacular kidnappings. But the figures speak for themselves: in 2006, over 14,000 people were victims of kidnappings.

The stories covered by the media do not even reflect the tip of the tip of the iceberg, and when they are mentioned, it is only because the kidnappings are politically motivated and thus particularly dramatic. This danger is often played down in mid-sized companies, although these are the very kinds of companies which are a driving force in globalisation.

Many of these kidnappings are only successful because people are ignorant of the risks on site, and this ignorance has natural consequences. Behaviour training and security drills should be a standard preventive tool for employees who have to work in troubled regions, regardless of whether they are expatriates or only briefly deployed.

Business travellers are exposed to a huge range of dangers, particularly including theft, robbery, loss, illness, accidents (including car accidents), kidnapping, being flown home and death. As stated above, many companies do not understand how expensive it will be if a worst-case scenario comes to pass. Major medical treatment overseas or bringing a seriously ill employee home can cost hundreds of thousands of euros.

Seen in this light, it is especially critical for the potential hazards at travel destinations to be carefully analysed so that the appropriate preventive measures and

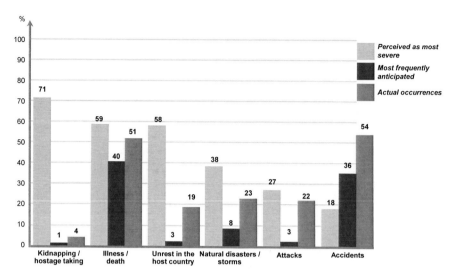

Fig. 1. Travel risks

crisis responses can be implemented. Nowadays factories are being built where there is access to resources. This means people do not travel to major cities, but to remote regions where not even local mobile networks work reliably. Risks like this must be offset, and the process needs to involve travel management. Arranging services to ensure the security of travelling employees should not be a one-man show run by the head of purchasing or an HR representative. Instead, everyone whose job description includes security in any way needs to sit down together to create proper crisis management, as opposed to travel management working alone. It is in this setting that the key information can be established, such as who is considered a frequent traveller, where they are to be sent, how long do they have to stay in a troubled region.

It is only in rare cases that travel management bears the sole responsibility for security-related processes and decisions that affect business trips. Incidents in recent months have led to a shift: an exchange of information is starting to occur between travel management and those in charge of safety and security in companies.

Despite this shift, some interesting questions remain: does travel management know what employees really think about security on business trips? To what extent do decision-makers and those directly affected, i.e. the travellers themselves, see things the same way? There may be issues that are seen in-house as been relatively unobjectionable — do travellers share this opinion? Are decision-makers aware of the ambivalence with which a traveller may depart for 6 months in Indonesia or a week in Sao Paulo or two nights in Kiev? If so, is this ambivalence taken into account when purchasing- and process-related decisions are made? Overall, are the potential risks of travels being properly assessed?

3   Discrepancies Between Employees and Employers – How Do Business Travellers Regard Their Own Security?

In autumn 2007, the second part of the aforementioned security barometer study was conducted. It surveyed the relevant employees at the same companies to determine the extent to which travel managers' views of things conform to or diverge from those of business travellers.

Chronologically speaking, the second part of the study took place at a time in which two major developments were having a perceptible impact on mobility in companies.

First of all, an upswing in globalisation forces employers and employees to be increasingly flexible in their mobility. A mobile workforce is emerging so that companies can make sure they stay abreast of things on the international market. For employees, on the one hand this means greater professional potential and a corresponding rise in financial incentives. On the other hand, the new demands for more mobility also mean greater risks in terms of successfully arranging business trips and stays abroad. For example, the pre-departure preparation phase for overseas travel (especially to Asian boom markets) is growing shorter and shorter, and it sometimes disappears altogether if an employee has to manage a number of different appointments within, say, seven days and ten flights spanning three continents. Travellers in such situations are constantly forced to adapt to different climactic, political, medical / sanitary and cultural conditions. These burdens are more significant now then they were 10 – 15 years ago. It is of indispensible importance here that employers create in-house conditions which allow employees to manage these challenges. This falls under the obligations that employers have to keep their employees safe, and with regard to business travel, it is also an important element of corporate social responsibility.

At the same time, the war for talents appears to be gaining momentum. This means that it will be more and more difficult for companies to find qualified employees: competition over such people will increase, and there is a growing number of companies that offer Employee Assistance Programs to enhance their attractiveness as a potential employer. EAP's provide extensive services for workers and their families to compensate for professional and job-related mobility.

The question that crop up: how can these many needs of employers and employees be reconciled? Or rather, are there discrepancies in their mutual expectations, and if so, how are they addressed?

In an effort to find answers, business travellers were surveyed about how they perceived their security, the risks of travelling and how well they were protected by their employers and crisis management at their company. This year's study compares the results to last year's findings. One of the key results was that there are major deficits in terms of information and how well business travellers are prepared for potential hazards.

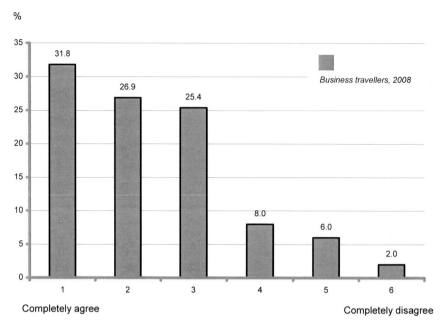

Fig. 2. "My employer is responsible for every aspect of my security on business trips"

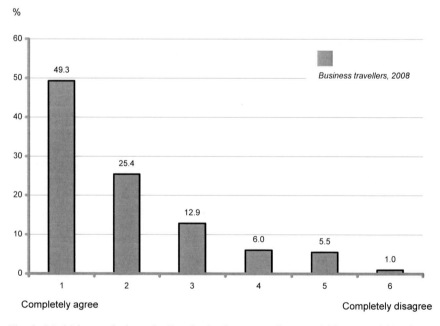

Fig. 3. "A 24-hour telephone hotline for business travellers would be a good idea in my opinion."

While 57% of business travellers stated that a colleague has been in serious danger at some time on at least one trip, travel managers assessed this risk as being somewhat lower at 50.5%. Just over one in five business travellers (21.5%) stated that they had felt left alone with their problems while en route. 87.6% think that a 24-hour hotline or a similar form of support for serious situations would be a good idea. If it improved their security, 94.1% would even approve of having their itineraries tracked by their travel agencies.

Another very interesting result was the fact that business travellers felt too poorly prepared for trips to troubled regions. Companies can respond by offering appropriate kinds of training which drill employees in the right behaviours during crisis situations. Ideally the course content could also be expanded to include intercultural topics. After all, what good does it do to arrive safely with pre-booked bodyguards if an employee then makes a cultural faux pas that dooms the long-anticipated trip (and the related expenses) to failure? The majority of surveyed business travellers wish they were given more general information about their destination (political situation, cultural differences, general security risk, medical and hygienic conditions). Only around a third of all companies give their employees this information, and a mere 11.4% train their employees before sending them into troubled zones. As a result, 85.1% of business travellers gather their own information — at least occasionally — about potential hazards.

The majority of business travellers (84.1%) also know about the legal obligations of their employers. 80.8% of the surveyed business travellers stated that they were insured by their companies against possible risks. The majority (85.6%) said

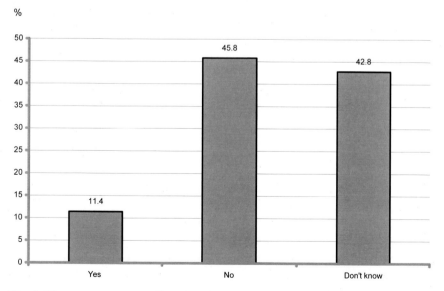

Fig. 4. "Does your company offer appropriate training for employees before sending them into troubled regions for longer periods?"

that they felt secure as a result. Many people, though, did not know whether or not or to what extent specific crisis-management measures would apply in emergency situations on business trips. Apparently there is a gap in information between travel managers and business travellers. In the event of political unrest, for example, 41% of travel managers stated that there were appropriate contingency plans. Only 16.3% of travellers, however, knew about these plans. In the event of an accident or death, 60% of travel managers were aware of contingency plans, compared with a mere 38.6% of business travellers.

What companies need is care management! And security is not the only issue that matters to business travellers on the road. Frequent travellers do not have time to take care of their everyday affairs, and as a result nearly half of them (47.2%) wish their employers would help support them. This correlates to the development mentioned earlier: an increase in mobility poses growing challenges to people's work-life balance.

Managers are becoming more and more sensitive to employees' security-related needs and support for frequent travellers / expatriates, because money and careers are no longer the top priority in many people's professional lives. It is important for employees to successfully integrate their private interests or harmonize their family lives with the demands of their work lives. In these days of advancing globalization, this trend is likely to continue. Companies need to prepare in a timely fashion by offering innovative "caring" services for employees; these enhance an employer's appeal, both inside and outside the company.

There is a simple economic principle that still applies, and it is something that is drilled into the head of every first-year economics student: "Cars don't buy cars!" The focus is back on people, since it is people (as employees) who make or break a company's success. A formula of sorts could be derived from this insight: Motivated employees = satisfied customers = profits = satisfied stockholders. Ultimately, everyone in the business process benefits.

4   Consequences for Travel Management

The in-house processes which affect business trips are the responsibility of the travel management department in larger companies and of office management (usually executive secretaries) in small and mid-sized companies. Due to the many cutbacks in service providers' commissions, especially at airlines, many travel management departments face the burden of having to justify their existence to the company. Among other things, this is taking place by means of travel management becoming more and more involved in the field of mobility management, i.e. assuming greater responsibility for all corporate travel. This entails new tasks such as fleet management or organizing employee deployments. Integrating security into the process of business travel is also an important aspect of corporate mobility, which makes it another vital area that travel management should cover. As a result, travel managers need to create the right kinds of networks to handle the issue of security on business trips. This includes a regular exchange of information among human resources, insurance purchasers, works security, company physicians and corporate security. All of these departments need to be heavily involved in establishing and implementing travel risk management.

**Beyond Tourism Industry's Boundaries:
The Philosophers' View on Sociological
Mega Trends**

The Collision of Civilisations and Challenges for the Global Tourist Industry

Asfa-Wossen Asserate

It is part of a travel agency's job to arouse curiosity and interest for other countries and other cultures. Travelling affords the opportunity to come into contact with people of different nationalities, to exchange ideas and to get to know foreign ways of life. Tourism can help promote the dialogue between people and cultures, and so promote an opposite position to the widely upheld view of a clash of cultures. The notion of a "*Clash of Civilizations*" developed into the popular catchphrase that it is today due to the book of the same title, written by the American political scientist Samuel Huntington. After the awful terrorist attacks of 9/11, Huntington was soon hailed as a prophet of his time. Today, his arguments help to justify the war on terror.

In summary, Huntington argues that mankind is divided into different cultural-political blocks, which he defines as cultural entities. He emphasizes the cultural fault lines between these cultural entities and assumes that, due to our profound and different values, we are almost forced to live together in a state of conflict.

In the tourism industry on the other hand, cosmopolitanism, open-mindedness and tolerance towards the foreign and unfamiliar are the prerequisites for success. It is not about the clash of cultures but the dialogue among cultures. Neither is cultural exchange a one-way street for tourism; rather it requires both sides to be open and ready to come together and learn from one another.

Hospitality belongs to the cornerstones of the travel industry, that is to say the openness and friendliness of the host towards their guests. On the other hand, the guests' respect for the cultural peculiarities and traditions of the host country is equally important. Understanding, openness and curiosity about other cultures and ways of life form the foundation of wanderlust. The tourism industry is undoubtedly one of the winners of globalisation, if one understands the latter as the countries of our world moving closer together. This can also be demonstrated by impressive statistics.

The *United Nations World Tourism Organization* (UNWTO) announces new record-breaking results every year. 842 million international tourist arrivals and tourist receipts of 586 billion Euros were the official figures for 2006. The year just gone once again saw considerable increases and, according to projections, it is likely that international tourist travel will almost have doubled once more by 2020.

The Germans have been world travel champions for years. Annually, Germans make more than 40 million trips abroad to countries all over the world. However, the Germans may soon lose their title. India, Russia and especially China are establishing themselves as the new giants in the travel industry. In 2007, 34 million Chinese took a trip abroad. According to UNWTO projections this figure will multiply in the years to come.

It is expected that in 12 years, the Chinese are going to take more than 100 million trips abroad. The curiosity about foreign countries, so long suppressed by the government, has created a longing stronger than we can imagine and money is no longer a problem. Approximately between 60 and 80 million Chinese could currently afford a trip to Europe, concludes Dr. Silvia Lot in her study, titled "What you have always wanted to know about Chinese tourists", carried out in conjunction with The German Chambers of Industry and Commerce.

India also has huge potential, with annual economic growth of around 8%. According to forecasts, by the middle of the century India will be the third largest economic power worldwide, after the USA and China. 300 million of India's total population of over a billion belong to the middle class and can afford to go on vacation. Indians are also greatly curious about foreign countries. The last four years have seen an average annual increase of outgoing Indian tourism of 15 to 20%. The consequences that the worldwide tourism boom will have on the world's climate and the holiday destinations are another topic altogether.

In any case, Europeans and Americans have to get used to the idea that in future they are going to meet increasing numbers of Asians in the world's most beautiful spots. Tour operators are already more and more adjusting their products to suit their new customers' needs and desires. This may irritate their regular customers, for there are bound to be differences in some of the tourists' customs. As the wise Chinese Confucius once said; "Men at their birth are naturally good. Their natures are much the same; their habits become widely different" (Translation by Herbert A. Giles).

In recent years, regular tourists from Germany, England and the Netherlands have been alarmed by media reports on the behaviour of Russian tourists. The fun-loving, spend-happy Russians are usually welcome in holiday destinations.

However, when Russian, German, English or Dutch guests meet at bars or salad buffets in the Mediterranean, this apparently does not always tend to go smoothly. "Russians have different eating and drinking habits, different tastes in music and different expectations of animation", admitted a speaker for the TUI subsidiary company Mostravel which nowadays sells holidays abroad to well over 200,000 Russian holiday-makers per year. Yet only caricaturists should be allowed to refer to a clash of cultures when speaking of alleged fights about deck chairs and lights-out times in hotels in connection with a clash of cultures. The issue is more of a lack of culture and at least of the most basic manners in holiday destinations. Natives from every country and region can tell stories of inadequate and excessive behaviour of guests.

Nowadays, many people around the world succumb to the enormous mental pressure the working environment projects onto their lives and they seem to find refuge in their leisure time. Especially on holiday, the most pleasant time of year, holiday-makers feel there ought to be no rules or regulations. During this time of leisure, manners of any kind seem to be nothing more than a burden, which we consider ourselves entitled to escape from. Thus, from time to time in holiday destinations, the most pronounced vulgarity reigns. However this vulgarity cannot be attributed to specific nations; it always emerges when people assume that they can do whatever they please. It can be traced in every society; it is a phenomenon which crosses boundaries of nations and cultural entities. Essentially, all civilisations and cultures are wondrous hybrids. They developed through exchange and synthesis, as well as the meeting of different peoples, religions and philosophies throughout history. At no time have various cultures been pure and closed entities. First of all, it might be wise to make a distinction between holidays where people want to escape from the constraints of their everyday lives for a while, staying in holiday-resorts for the masses; and trips on which they want to explore unknown places around the world for themselves. The former alternative demands holiday deals that are easy and fast to reproduce, standardised products, allowing tour operators to generate maximum economic profit and follow the logic of globalisation. This would usually include sun, sand, and sea as well as a rich selection of food and lots to drink. Above all, tourists are not to feel overly foreign anywhere in the world, and the surroundings are jazzed up to perfection. Experience has shown that, all over the world, nothing sells better than a streamlined product. But of course, no one would expect this kind of tourism to lead to a more intense or more dynamic exchange of cultures.

As in other areas, this capitalist take on tourism has a negative effect on diversity. Local cultures and alternative ways of life become marginalized and even extinct.

Travelling, however, particularly requires an interest in things that are different to home. The prerequisites for this mind-set include a certain freedom from self-indulgent dogmas, a fundamental curiosity and intellectual tolerance.

If nothing else, it demands an interest that goes above and beyond the sheer pursuit of profit and personal gain. Here, the challenge for tour operators might involve attempting to generate interest and curiosity about places beyond the safety of the walls of holiday resorts. Nowadays this holds especially true for travels to Islamic countries. Astonishingly, it is tourism to precisely these types of countries that is observing an enormous rise in demand and growth; apparently in defiance of the difficult political climate. Even package holiday tourists can discover the true reality – the one that lies somewhere between those polished catalogue pictures and the horrifying images in the news. All they need is a little curiosity and an open mind.

Traditionally, encounters with the country and its people are the focus of study trips. Trips to Teheran or Damascus or even journeys to the Bedouins in the desert, undertaken by small groups of tourists, may even amount to mutual understanding among nations on a smaller, more personal level. For the rural population in some countries, encounters with tourists are often the only possibility to come into contact with the Western world. Tourists, on the other hand, are able to discover first-hand that the reality of Islam differs from how it is represented by fundamentalists.

Most Chinese tourists wish and hope to experience Europe for themselves. The number of Chinese tourists to be seen photographing typical European landmarks such as the Eiffel Tower, the Brandenburg Gate or Neuschwanstein Castle is fast increasing. Apparently, like China's rapid economic boom, the first trip to Europe ought to be as swift as possible and leave nothing out.

In travel agencies, most Chinese tourists choose typical packages for their trip to Europe, which offer tours of five countries in eight days for 750 Euros, or 12 countries in 16 days for 1750 Euros, whereby they usually travel in larger groups. Thus, Chinese tourists spend most of their "Tour de force" on motorways, becoming acquainted with Europe mainly through the windows of their busses. So it is hardly surprising that Chinese tourists, by the end of their travels, find Europe disappointing and overrated. Many come with false conceptions, and consequently when they arrive in Europe's industrial countries they are irritated by the lack of skyscrapers and the abundance of green countryside. It seems that even this fast and modern way of travelling fits in with the deduction made by the English statesman and humanist Thomas More on Christian pilgrims as early as the 15th century, later concurred by the French Romantic writer Chateaubriand; "There never was a pilgrim who did not come back to his village with one less prejudice and one more idea."

Yet, at the thought of the speed of travel of Chinese tourists, a quote by Goethe comes to mind: "You haven't really been somewhere unless you've been there on foot."

However, we should beware of viewing China in a disdainful way. Today, Germany has a population of about 80 million people whereas China has a population of approximately 1.3 billion. China ascended the economic ladder in an incredibly short period of time and is the world's fourth-largest economy. Every year, the economy accrues approximately 8.0 percent and Chinese national pride grows with it. Traditionally, China calls itself the 'Middle Kingdom'. And for many centuries China genuinely remained undisputed in the spotlight, at least in the Asian world. For around 150 years, up until the end of World War II, China was attacked and exploited by European powers and Japan. The proud country was deeply humiliated. This remains a deep wound to this day, and almost certainly acts as motivation for the incredible economic boom, that China has achieved in recent times.

Thus, the collective target is to regain the lead in the international community. Kai Strittmacher, former Beijing correspondent for the *Süddeutsche Zeitung* and author of a practical guide for China, suggests that there are souls hidden behind the new wealth disoriented at the breathtaking changes in their country, with skyscrapers, internet, and even MTV. However, one can sense a feeling of emptiness. Behind the curtain, there is a sort of yearning, a hunger, which has yet to come to light. This is due to the fact that, before becoming an economic power, Communist China smashed its own cultural history. Mao Zedong and his followers committed themselves to the theories of a certain lawyer from the German city of Trier, whose grave remains a site of pilgrimage for Chinese tourists. Nowadays however, China has joined the international ranks of market economies and demands a leading role. Prof. Dr. Wolfgang Arlt, head of the "China Outbound Project" at the Stralsund University of Applied Sciences, interprets the Chinese tourists' interest in western industrial countries as part of a search for orientation. He states that "the Chinese want to find out how much catching-up China has to do in order to reach the peak." In fact, many Chinese actually believe that China, as the cradle of civilised culture, is on the path to becoming the global power it once was.

One could argue that this again refers to Samuel Huntington's theories about the inevitable *Clash of Civilizations*. In any case, most Europeans believe Europe to be the centre of the world and that any other belief is a threat to world order. The last five hundred years seem to confirm this point. Without Europe, there would have been no Renaissance, no Enlightenment, no French Revolution, no Industrial Revolution and no Modernism. Without a doubt, Europe has achieved a lot. The notable success of the French and British colonial empires, as well as the spread of the scientific systems perfected in Paris, Berlin, Vienna, London and Rome, mislead people into thinking that the world's driving forces have always originated from the West.

Even today's balance of power supports this view to a large extent. European powers not only conquered most of the world; they also exported their languages to foreign countries and their educational and administrative systems all over the globe, including even the United States of America, although the Americans like to stress their uniqueness. European values seem to be universal nowadays. Everything that is not in line with these values is regarded as regional. The question is; what exactly does European mean?

Many people base their identities on the apparent stability and uniqueness of European cultures. These identities are often portrayed as static and innate. The disparity to the other, to that which is unknown, is defined as irremovable and irreconcilable. Yet whoever clings to this opinion disregards history.

In Greek mythology, the home-town of the beautiful Europa was a trade centre on the banks of the river Euphrates in today's Iraq. There are many different versions of the story of Europa, the Phoenician princess. The settings and plots fluctuate and the moral and political messages vary. This can also be applied to Europe as a continent, since it has always been a junction of various cultures and streams of ideas, as Ilja Trojanow's and Ranjit Hoskoté's book *The Clash Denial – How Confluence saves Civilizations* so nicely demonstrates.

The following example may demonstrate how not only ideas but even objects, which are today considered to be genuinely European, are in fact imported from other cultures. Over many centuries, the Byzantine court was well-known for its elegance and etiquette. In the thirteenth century, Dominicus Zelwo, a nobleman from

Venice, chose a Byzantine princess to be his bride. When the princess used a fork at the wedding banquet, she created a scandal. In reaction, the Cardinal Bishop Petrus Damiani is said to have uttered the following words; "God in his wisdom provided man with natural forks – the hands. Therefore it is an insult to Him to substitute artificial metal forks for them when eating." This clerical denunciation banished the fork again to the cupboards of European history. It was not until three hundred years later that it became fashionable in Italy to bring one's own fork and spoon along to banquets, which were elegantly stored in a case called a "cadena". The rest of Europe still largely rejected the use of cutlery. Apparently marriages across territorial boundaries often afford cultural exchange. Since, when Catherine de' Medici married Henry II of France in 1533, her dowry also included silver forks. However, the French royal court continued to eye these novelties suspiciously. Even Louis XIV still preferred the use of a knife and his hands.

In contrast, in the East, in the Byzantine court, forks had been used since the fourth century. A letter by a Franciscan monk to Louis IX of France shows that even the Tartars were experienced in using forks. Nowadays, the fork is regarded as an important European achievement, and people who still use their hands are at best considered charmingly curious. There are countless items that were imported from other cultures and are now classified as genuinely European, such as perfumes, libraries, or even coffeehouses. Objects were assumed in the same way as stories and ideas. Initially, they are treated with suspicion followed by preliminary acceptance. Then, what was once alien is enthusiastically taken over, until its foreign origin is forgotten. Being myself Ethiopian, I do not want to miss the opportunity to praise the classic way of eating, namely eating with one's hands.

In the meantime, this custom has become synonymous with barbarianism in modern Europe. Few have actually witnessed this convention in countries and places where it is still practised, such as in Turkey, India, African and Arabic countries. In fact, it takes more dexterity and skill to eat with one's hands elegantly and gracefully than you might assume. After all, it really is much more difficult to eat with one's hands than with knives and forks. Of course, washing one's hands belongs to the meal. In the countries mentioned, large jugs of warm water are passed around before and after meals. In wealthy houses in India, a bowl filled with sprigs of laurel stands at the ready. The water is then poured over them, so that nobody has to see the used water.

This was the way one ate in the ancient world, and even the Germans used their hands up until to the middle of the 17th century. Incidentally, recent research has shown that a certain enzyme is released on the surface of our fingers that aids digestion. I for one would recommend consciously taking the opportunity during your travels to give eating with one's hands a try. That way, the ancient European custom might return to the Western world.

This short digression was meant to demonstrate how cultural traditions adapt and change and how much they are influenced by other foreign traditions. Is it not true that in any major European city, anyone who cannot eat with chopsticks is considered to be a red-neck? Where would we be without the benefits of Chinese medicine or acupuncture? Not only cultures change constantly; we all do. We are all dynamic people and we are constantly learning. Every single one of us is a cultural hybrid; influenced by cultural traditions and customs whose true origins mostly remain obscure. Every human and all cultures are invisibly interconnected.

In Indian tradition, this is beautifully described with Indra's net. This old Buddhist God is pictured with a net that stretches out infinitely; every node of the net being adorned with a jewel. Each jewel represents a single being and each of these reflects the rest. If we think of ourselves as one of Indra's jewels, we will not only see the individual soul in a particular body but we will also see all the others we are connected to. All those other jewels reflect back on us and enrich our lives. To most people's amazement, nowadays even physicists resort to similar images, when they try to describe the genuine essence of our world. Everything is interconnected and interwoven. All our actions reflect back upon us and have unforeseen effects. Considering all of this, the idea of clash of civilisations seems utterly absurd. Every great civilisation draws its wealth from the merging of many cultural traditions.

Deep down we all know this, and the travel industry survives off of the resulting deep-rooted curiosity about the 'other'. This merging of cultures is dependent on the mobility of people, ideas, goods and services, towards which the tourism industry makes a significant contribution.

In our turbulent times, cultural diversity and cosmopolitanism are virtually necessary requirements of human existence. The peaceful coexistence of cultures and mutual appreciation are more than essential to our lives today. Anyone who sets about achieving this soon recognises that the other party is no enemy, no alien, but basically just a mirror of the various possible facets of ourselves.

We should look into this mirror, not to loose ourselves in confusion, but in order to recognise ourselves and our possibilities. Success in tourism is heavily reliant on a stable international situation. It should therefore lie in the industry's very own interests to spread this message to every last corner of the earth that cultures don't conflict with one another, but merge. And we should defend ourselves as a group against all those, who try to alienate us from each other in the name of disparity.

Tractatus Philosophico-Touristicus

Peter Sloterdijk

What I would like to talk to you about really cannot be called anything other than what it says on the title, i.e. *tractatus philosophico-touristicus*. In the paragraphs of the article, I have provided outlooks of the contemporary economic and tourist world scene about which I hope to be able to talk to you in the remaining available time.

I start with an observation, which I'm sure we all share, that world tourism must be considered one of the key factors of the current as well as future global economy, more still in actual fact, of global culture, of global civilization. Together with the all-pervasive and – for the time being – all-enabling market for fossil fuels, tourism, say experts, is actually and potentially already the biggest industry in the world's spectrum of economic activity. And these two economic systems are highly enmeshed with each other with regard to the new mega-projects, for it is readily apparent that 21st century tourism will remain engaged in the end-game of fossil energy.

Mass tourism is essentially long-distance tourism. And long-distance tourism relies on aeronautical systems which in turn depend on fossil fuels for the time being. We are, in a manner of speaking, in a self-justifying circle from which exit is unpredictable. By the way, there was perfidious and undemocratic but yet realistic speculation in Pentagon circles at one time, to reserve the world's oil output of a certain year using military means for aviation, so to say, to block civilian use of petroleum reserves so as to secure the option of aeronautical sovereignty. That is speculation of a hybrid nature but one which permits us to dream about the chances of global mass tourism which cannot be organised only as global mass tourism of millionaires. Indeed, there are so many millionaires – this is perhaps the most important news of our era. Today there are so millionaires as there were ethnic groups in the past. There is a meta-nation of millionaires, there is a new kind of wealth which meanwhile is so disengaged from the lives of others, as a result of which the millionaires do not really believe that they are involved in the endgame of fossil energy. This is the premise for what we shall discuss today under the heading of Mega-Destination under the assumption that these Mega-Destinations represent at the same time "Hydro-Projects" in which only invited buyers participate. Among this new population, this meta-population of millionaires and billionaires, there are also plebs who are not invited. This is news that should be music in the ears of the old proletariat because it offers or the first time ever, the possibility of rubbing shoulders with shunned millionaires.

With the entry of former emerging economies into the open world market especially with the advancement of new global players like India and China and the attendant emergence of consumerist middle classes in these as well as in many other countries, a vast reservoir of customers is emerging, who already fulfil or will soon fulfil several prerequisites to participate in the global mobility market. These prerequisites are surplus purchasing power, disposable free time, ambition for status, need for regeneration and a form of curiosity that is at once generalised and specialised, which I would like to call *conditional xenophilia* further down. I think that is, in a manner of speaking, the anthropological basis for what is called tourism. This was the pattern on which the tourist habitus was also built in the past and will probably continue to do so in future.

In its mass, organised and global form, tourism is a phenomenon which systematically depends on the technical and socio-psychological framework conditions which have been created in the course of past centuries. If one can claim with good reason that 20th century civilisation can be attributed, for better or worse, to the largesse of cheap fossil energies, then this is all the more true for one phenomenon of ultimate luxury of our era – namely the first-time, singular integration of broad sections of society into tourist practices, which – not coincidentally – were the prerogative of aristocracy and the *grand bourgeoisie* in the past. So no Goethe without Italian Journey, no 19th century person of letters without a similar passage of initiation for the sake of art.

Only out of the encounter between the seemingly limitless disposable energies and new motorised mass-transport machines of the 20th century could emerge a *kinetic democracy*, a term that has been used to describe the motorised society or simply the mobilised society.

What we discuss today on tourism is by and large a metamorphosis of kinetic democracy. Mass mobility, as I said earlier, is today aeronautical. Flying too, is a democratic form of movement. It belongs in the era of cheap energy and will progressively be drawn into the era of shortages and rising prices. This situation will by far not be as dramatic as long as oil prices do not exceed a threshold of, say, 300-400 dollars per barrel.

From an anthropological standpoint, tourism can be classified under release phenomena. Human modes of behaviour could be considered release phenomena if they do not demand complete seriousness and ultimate commitment from the human subject. Art is thus considered a paradigm for released behaviour as it helps human beings to cross into a second world, a universe of simulations where the conditions and constraints of the first reality do not apply. The lack of seriousness should not be understood as a deficiency or flaw. Friedrich Schiller in fact claimed ultimate existential value for such released situations – true to the well known dictum that a person is a consummate human being only at play. This thesis can be applied *mutatis mutandis* also to people on journeys, for a person is a consummate tourist only at play with the journey – that means that the person is released from the unreasonable demand to justify the fact of the journey from A to B with serious explanations. Let us state for the sake of definition: any person forced to take travel seriously cannot be a tourist. My commiserations to all those people who work in the tourism industry, to those who cannot enjoy any release and give themselves up to the pleasures of travel as is characteristic for the true tourist. The coloniser, conqueror, missionary and explorer, the research traveller, the business traveller, the seaman, why, even the tour guide – and I am sorry to say this – are all fake doubles of tourists because their movements in free space are motivated by serious purposes.

Europeans, who should perhaps be considered inventors of tourism, did not arrive upon the release and aesthetisation of travelling from one day to the next. The emancipation of travel from economic, political and scientific purposes itself had to come a long way. Prototypes of tourist behaviour could even be discovered in ancient Greece and Rome for instance, in the discovery of cool summer resorts, of Artura village, by affluent and genteel circles. These circles believed in any case that human existence begins where labour ends, above all that being a human, true *humanitas* begins when one resides at one's *tusculum* in the countryside.

These early notions of tourism also appear in the fostering of the idyllic as a poetic and perceptional form in late antiquity – for *eidyllion*, literally the 'small picture' is nothing other than the verbal form of a flyer for what urban folk even then considered a more attractive, simpler, better and more rural life. The Greeks, of course invented the postcard and that in the form of a verbal postcard, called *eidyllion*, idyll; an idyll is sent to others with greetings from the holiday. Furthermore, medieval pilgrimages could be considered precursors to tourist mobility although the purported holy purpose obscures the playful evasion.

I would not like to withhold from you the definition of 'Tourist' in the German *Brockhaus Conversationslexikon* of 1855 because it still captures the essential idea very well. The entry reads: designation for a traveller who does not link the journey with any e.g. scientific purpose but only travels to have made the journey and be able to describe it subsequently.

In the meanwhile, the relationship between travel and its description has been significantly dissociated and travellers of the post-bourgeois era conduct themselves also in a post-literary manner as a rule, only in exceptional cases do they view the journey they have undertaken as part of an educational process. Nonetheless, the genre of souvenirs and mementos has not suffered as a result of the change, quite to the contrary: ever since easy-to-use portable cameras are available, private documentation of journeys has grown into an immense market, and whatever may have been lost in the realm of literary culture has been more than compensated by gains in the realm of visual culture. There is, after all, a strong symbiosis between tourism and photography. This is a market which also needs to be studied in detail: the mobilisation of images which is one of the principal characteristics of present day culture, is a side aspect of mass tourism.

What I want to say with the above is that between the 17th and 20th centuries, aristocratic and elite travellers established a habitus that could become everyman's patrimony in the 20th century. What we are observing in nothing less than the democratisation of luxury: being what it is, tourism as one of the most powerful trends in modern economic society at large enables the transition from elitist consumption to mass consumption, the transformation of historical privileges into objects of common use. In the meanwhile, we can hardly remember the times when having been abroad meant a more or less sensational exception for the large majority of the population. We recall how Othello, Moor of Venice conquers the heart of a young Venetian not by virtue of his physical attributes but by narrating how he suffered abroad, how he was almost devoured by cannibals among the Moors. So moved was the heart of the young Venetian woman by his words that she could not but love him. "This to hear would Desdemona seriously incline", says Othello.

So travel, ladies and gentlemen, and come back with the Siegfried or Caribbean factor and you'll soon be winning!

In this age, travel to foreign countries has become a fact of life for the large majority. This has significant political implications not least for the expanding European Union. While that is not our topic today, it should be clear that Mega-Project and the Mega-Destination of all Mega-Destinations is Europe. We must entice Europeans within Europe to convince themselves that Europe exists. I believe that for Europeans the Mega-Destination of the future is Europe, for we have 27 destinations within Europe in the meanwhile of which we do not know most and we must convince ourselves of who our neighbours are, who our fellow-citizens are – the effort is worth every investment.

The formula for the democratisation of luxury is obviously paradoxical. It says on the one hand, that needs that counted as luxury in the past could be transformed into basic needs in the course of social evolution; on the other hand, it suggests that in the consumer society, the gap between what is necessary and what is satisfying is widening – which leads us to conclude that satisfaction drops by the extent to which the means for achieving it are popularised. As a consequence, the lives of the broad middle classes – as only those of aristocracies in the past – are falling increasingly under the law of rising demands and intensifying stimuli. These transformations would not be possible in the natural course of things if modern mass consumption were not backed by a powerful overall economy capable of providing large section of its participants with good opportunities for mobility, sufficient surplus purchasing power, generous holiday provisions – also an important theme – and comprehensive educational opportunities, not least, training in foreign languages. Rising life expectancy also points in the same direction. Certainly, popularisation of luxury motivates the luxury of the elite classes to establish higher levels of distinction. It is a well known fact that the luxury effect is not solely a result of material content but in equal measure and more from exclusivity. In the era of mass tourism, this means that in the upper wealth segments a clear upward distinction must take place. This is what in the course of the recent years and decades, has given rise to a flourishing upper-crust hotel industry, which is unprecedented in the history of mankind in terms of scope and interconnectedness. In view of these phenomena, it may be asserted that the much-lauded democratic elitism has quietly seeped across the threshold of neo-aristocratic forms of life. This is where true esotericism or rather the esotericism of prices has paved the way, beside the visible movement of the masses, for a discreet world of locations for highly individualised luxury travel activities to evolve. And the Mega-Projects of today and tomorrow are founded just on this expectation, that in this segment, in this luxury segment, a new neo-aristocratic sub-culture will be established. Do not forget: neo-aristocracy also signifies the return of what has been eradicated in the 200-year history of the emancipation of the citizenry, that is, of the serving personnel. 'Personal Computer' could also be taken to mean that people who do not employ any staff[1] may personally serve their machines.

In this regard, the German language is in any case the most philosophical and the most sociological of languages because as German speakers, we have decided to say from the very beginning that we serve[2] machines. This a pointer to the man-machine complex from which there is much to learn.

Tourism is not just a social phenomenon, it also extends a little into the anthropological realms. There is always in humans something like xenophobic but also xenophilic, neophilic tendencies. Tourism if you like, is applied neophilia i.e. applied curiosity, the joy and love for everything new. Of course tourism also has another branch to it – I'm sure you have already seen it. There are many people, who travel only to conclude that elsewhere is worse than at home and return enriched by this insight. This is of great importance to the realignment of the immune systems of human beings in modern society. I do not unfortunately have the time to expand on these edifying thoughts in detail. They do belong, in a certain sense, under the heading of 'Democratisation of Luxury' insofar as the luxury of mobility, the luxury of disenchantment, the luxury of negativity can at all be drawn in to represent the essence of luxury. If one has got not just one but several different illnesses concurrently, then one is indeed over the worst, we may say.

From a broader cultural-anthropological perspective, tourism in our era represents a profound modification of human styles of existence, which we may summarise as the disintegration of *monolocal* ways of life. This wording gives a more precise idea of processes that may be found in the literature under terms such as *new nomadism* or *departure from sedentariness*. The transition from an agrarian world to ways of life in an industrialised society, i.e. urban ways of life, has very often been described as the most encompassing, civilisatory drama of our era. And this is very important: for our context, this means that the transition from post-agrarian ways of life was in fact also linked in most cases, to a transition into *plurilocal* forms of living and working which has resulted in micro-mobility on an immense scale – it is commonly called commuting. Drawing an analogy between the pre-agrarian, nomadic ways of life and post-agrarian 'neo-nomadic' lifestyles is in a sense quite justified, if we consider the fact that the real nomads, pastoral

---

[1] Play on words: *Personal* in German means 'staff' or 'service personnel'.

[2] Play on words: the verb *bedienen* in German means 'serve' as well as 'operate'.

peoples, were as a rule not ceaseless wanderers as romantic notions lead us to believe, but rather commuters who oscillated between the summer and winter pastures of their herds. This is the famous phenomenon of transhumance to which is linked the movement of humans. And we can use this to make a distinction between archaic and modern commuters in their oscillation between the summer and winter abodes. The latter group does not follow the availability of favourable fodder for its herds but rather favourable climate for itself. Perhaps the starkest innovation of the modern age is that human beings do not view themselves as creatures of the clime – *klima* in Latin-Greek, i.e. of the angle to the heavens under which they are born, but that they have started to capture for themselves a new type of freedom of climatic choice. This is absolute blasphemy, for you know that fate and weather is God-given and anyone who does not accept that is an atheist. This type of climatic atheism has taken deep roots in us in the meanwhile, for who among us has not cursed the weather, who among us has not drawn due consequences from the cursing and deserted? Think of what happens in Germany at Christmas – that is nothing but mass desertion from a common fate which increasing numbers of people believe they can escape – by just packing up and leaving the Nation, the Bad Weather Community.

Monolocalism in its present-day form means not as much persistent residues of agrarian ways of life but is rather an indicator of poverty, of social degeneration or immobility due to illness and age. If on top, one considers the traffic-sociological statement that only one out of three instances of traffic movement is work-related – and this is important to realise, I believe – and that the second and third instances of traffic movement belong to the growing universe of lifestyle-mobility, it is then that we begin to comprehend the sheer scope and significance of this subjectivised, luxury-tinged traffic culture. Lifestyle mobility includes weekend excursions, shopping trips, having to take one's daughter to ballet class, going to the theatre, infidelity in conjunction with journeys etc.

The Mega-Projects of today and tomorrow presuppose a clientele for whom bilocalism, trilocalism has already become second nature. One might say, a new species of humans is evolving, to pick up on the concept of a hybrid species again, a species of humans disloyal to any place, a species that has elevated infidelity into a way of life to systematically cheat on one house with another, on one apartment with the next and on one island with another island.

I must come to the conclusion. I shall close with an idea from cultural-ecology. It is a well-known fact that physical global transport over the last 500 years has opened the world up to create a networked complex of transport, information and economy. This is what the word 'globalisation' really means. In this respect, one may say that traffic is fate – and this in a dual sense – as physical transport in which bodies are moved as well as mental traffic in which data and emotions circulate. Now, the unexpected side-effect of long-distance movement and long-distance view is that the conventional system of distances between countries, cultures and people breaks down – with all the consequences that this transformation implies to the world maps. In the mental sphere, physically remote elements have become immediate neighbours. Imagine, that a year ago an unexpected neighbourhood between Denmark and Syria came into existence or that USA and Saudi-Arabia have been adjacent to each other for quite a while on these new world maps of infections, maps of irritations – these are what represent the true geography of our time. And when one knows – and this is not an insight to be underestimated – when one knows, that neighbourhood and enmity are a habitual pair – this can be construed from lexicons in almost all languages of the world without much effort – then one also knows that the consequences of globalisation cannot be harmless. The recent frictions arising from the cartoon dispute have demonstrated this explicitly. Modern telecommunicative traffic has given rise to new categories of collision, collision in symbolic space, which were entirely unknown in the ancient and medieval world. They belong to the class of civilisatory phenomena called 'harmful density'. This is a phrase that I introduced in one of my earlier books to describe how density phenomena transpire in the modern world which subject human beings to greater civilisatory stress. Density, in a manner of speaking, means the increase of collision probability or put more neutrally, of encounter probability and this must be made civilised as a whole. The process of civilisation – this is an idea that Norbert Elias brought into the debate several decades ago – is among the most fertile in contemporary culture theory. The process of civilisation is essentially the process of making oneself dependent on somewhat remote neighbours and by the same process, of becoming denser, raising the probability of encounter, that requires something like a code to ensure that every unexpected encounter does not also become the collision which the encounter embodies. This means that the harmful density must be computed if one wants to draw up a fair balance sheet of the costs and benefits of modernisation. This implies such phenomena like tele-competition, tele-denunciation – now that would be a fine concept to hold a media-theory or journalism seminar on! We know categories of tele-criminality, we also know tele-war and similar forms. Tele-eroticism is also discussed of late. I have even read of tele-dildonics in an important textbook on cyber-reality but I am not going to explain that so as to not give you bad ideas. Similar sinister tele-consequences are reported from all quarters. In the eco-systemic view, more communication does not mean just more harmony and chances of harmony but also more instances of conflict between spatially remote partners in the modern world. Globalisation means a modern form of the world in which the classic, spatially hygienic distance collapses. Since we are transported into a system of remote neighbourhoods, we stand adjacent to this, that and everything else and so we must expend extra effort to restore the old distances on our psycho-mental map of the world or even define new distances. The vision of a global village of media prophet Marshall MacLuhan does prove to have a dark side, which the great thinker came to see only in the latter phase of his work on media theory. The global village not only means that the world, synchronised by electronic media, will be pacified by some kind of an evangelical tam-tam. MacLuhan made the global village a pivotal element in media theory because he believed that we can, in a manner of speaking, technically imitate the Holy Ghost using the distance-shrinking effects of telecommunications. The world would become something like one large Pentecostal service, a Pentecostal event and the old village tam-tam would turn into a world tam-tam. We could allow ourselves something that is essential for harmonious communication viz. the transition from verbal communication (which always finally ends up in quarrel) to musical communication where rhythm synchronises us all and creates something that we call harmony. Even that is not quite right, in the final analysis it is rhythm and not a common melody that synchronises human beings because melody too, is something we can quarrel about. Above all one might sing off key and that may not go down well with the neighbour. You do see how the German footballers sing the German national anthem. That should be reason enough to stop the game. Fine, but I do not want to go into that in greater detail.

This context allows us to present our concluding thesis plausibly, which is: mass tourism in future will increasingly become a political issue. I believe that the complex is also reflected in this event and it must become a political issue because the primary topic of the 21st century, after all, is security. I think we've all understood that. We have manifestly left behind the era of freedom, the overarching topic of freedom which has dominated the last 200 years. For freedom, to put it maliciously, is really something that interests only tax evaders while all other categories in society have switched to this Mega-Theme of the 21st century – security. And that is something that is of great importance for future tourist activities because even practices relevant to the civil society such as travelling need to be reviewed afresh under the aspect of the security-imperative. It is as good as absolutely certain that this cannot happen without sacrificing freedoms. One should bear in mind from the outset that very many of these Mega-Destinations which one talks about, imply political moves towards a kind of apartheid society, steps into a society of gated communities and luxury tourism, which by its very nature, is not possible without such division and insulation.

Since freedoms are not operable without security, even the most resolute of the liberals will not get away without making concessions. The foreseeable conflicts will prove true the definition that politics is the art of the possible. I hope I will not frighten you by using a few philosophical terms, having tried so far to skirt the terminology of my subject as far as possible: from a philosophical perspective, I should add that the sense of possibility does not lie, as Leibniz says, in pushing an individual value to the extreme but rather in taking several or many concurrently observed values to their respective optima. That means, we should offset one exaggeration by others because several exaggerations should exist beside each other. Leibniz calls the parallelogram of several relative optima *composability*. I think that is one of the words that we should include in our vocabulary for the 21st century. Composability – that means the making possible of things that tend to mutually exclude each other. Politics being the art of multilateral possibilities is – here comes the terrible word – applied *compossibilism*. Would you please note it down, compossibilism. It signifies the art of arriving at the least bad result from conflicting values in suboptimal situations. And when a maximum of unilateralists is unhappy, that could prove that the art of the possible is operating close to the optimum.

Where Global Trends are Revealed.

ITB
BERLIN
CONVENTION
THE LEADING
TRAVEL INDUSTRY
THINK TANK

11–14 MARCH 2009

Official Partner Region
ITB Berlin 2009

RUHR.2010
European Capital of Culture

tourism metropoleruhr

Co-hosted by

Messe Berlin GmbH · Messedamm 22 · 14055 Berlin · Germany
Tel. +49(0)30/3038-0 · Fax +49(0)30/3038-2113
www.itb-convention.com · itb@messe-berlin.de

IIIIII Messe Berlin

Printed by Books on Demand, Germany